THE CLINICALLY ORIENTED GROSS ANATOMY LAB WORKBOOK

Frank J. Slaby, Ph.D.

Professor

Department of Anatomy & Regenerative Biology

George Washington University School of Medicine

Washington, D.C.

ISBN 1453650709
EAN-13 9781453650707

DEDICATION

The author dedicates this book to his wife, Susan K. McCune, M.D., and our son, Christopher. Their support and assistance lightened and speeded the effort.

DISCLAIMER

The author has attempted to provide the most accurate information regarding the clinical application of gross anatomical knowledge acquired in the gross anatomy lab. However, the author and the publisher are not responsible for errors or omissions and in no way should the information provided in this book be applied to diagnose or treat patients or substitute for modern clinical training. The author and publisher make no warranty, expressed or implied, with respect to the accuracy or application of the information contained in this book in the diagnosis and treatment of patients.

HOW TO USE THIS LAB WORKBOOK

The gross anatomy lab is the only place where a student who is studying to be a physician, physician assistant, nurse, or physical therapist can actually see and learn how the organs, glands, blood vessels, nerves, bones, joints, and muscles of the body are spatially related to each other. The gross anatomy lab experience, however, is too often primarily focused on the dissection and identification of organs, glands, muscles, nerves, and blood vessels. There are not any gross lab dissectors or workbooks that associate the knowledge learned in gross lab with the skills applied in clinical practice. This workbook has been designed specifically to extend the gross anatomy lab experience into the clinical realm.

Each chapter in this workbook focuses on one of the major body regions. In the first pages of each chapter, questions of various types (multiple choice, fill-in-the-blanks, and essay) are presented to identify the clinical anatomy you should know about the body region you are dissecting in gross lab. The questions focus on the anatomical basis of common injuries, conditions, and diseases and the surface anatomy knowledge that is applied during physical examination of patients. The answers to these questions, which are provided in the last pages of each chapter, are what your clinical instructors will expect you to understand and be able to apply when you are in your clinical rotations. This workbook will help you experience gross lab in its most meaningful sense: the opportunity to have the cadaver, the physical remains of your first patient, help teach you how to visualize in your mind and examine the internal organs, muscles, nerves, and blood vessels of all the living patients you will try to help in your professional life.

TABLE OF CONTENTS

SHOULDER - Part A: Questions

Dissection of the shoulder in gross lab focuses on identification of the 15 shoulder muscles and some of the nerves that innervate these muscles. Three topics dominate the shoulder anatomy most frequently applied in clinical practice: (1) the nature of the three most common skeletal injuries of the shoulder, (2) the mechanism by which rotator cuff tendons commonly become inflamed or torn, and (3) knowledge of the prime movers of shoulder and arm movements.

Bones & Joints of the Shoulder

The clavicle, scapula, and the proximal, or upper, end of the humerus form the skeletal framework of the shoulder. The clavicle, scapula, and humerus form three synovial joints in the shoulder: the sternoclavicular, acromioclavicular, and shoulder joints. The three most common skeletal injuries of the shoulder are clavicular fractures, shoulder separations, and shoulder dislocations.

1. The clavicle is palpable along _____ in the upper anterior border of the shoulder.
    _____ only the medial third of its length
    _____ only the middle third of its length
    _____ only the lateral third of its length
    _____ only the medial and middle thirds of its length
    _____ only the middle and lateral thirds of its length
    _____ its entire length

2. The five most prominent parts of the scapula are its acromion, coracoid process, glenoid cavity, spine, and inferior angle. Which of these parts are palpable during a physical exam of the shoulder?
    _____ Acromion
    _____ Coracoid process
    _____ Glenoid cavity
    _____ Spine of the scapula
    _____ Inferior angle of the scapula

3. Explain why the clavicle is the most commonly fractured bone of the body.

4. Describe the nature of the injuries and the radiographic criteria of grade I, grade II, and grade III shoulder separations (shoulder separations are graded by comparing AP shoulder radiographs of the patient's injured and non-injured shoulders).

5. The shoulder joint provides the greatest range and freedom of movement of any joint in the body. List the three anatomical factors that significantly enhance the range and freedom of movement of the arm in the shoulder joint.

6. List the two factors that significantly contribute to the fact that the shoulder joint is one of the most commonly dislocated joints of the body.

7. There are two types of shoulder dislocations: anterior and posterior. What is the approximate percentage of shoulder dislocations that are anterior?

8. Explain why the axillary nerve is the major nerve of the upper limb most at risk of injury from an anterior shoulder dislocation.

9. Explain why the axillary nerve is the major nerve of the upper limb most at risk of injury from a fracture of the surgical neck of the humerus.

10. Abducting the arm 180° from the anatomical position involves coordinated movements at the sternoclavicular and shoulder joints. Describe the movements that occur at each joint during (a) abduction from 0 to 90° and (b) abduction from 90 to 180°.

Muscles of the Shoulder

The 15 shoulder muscles are listed here from A to O. Muscles A-D are the four shoulder muscles that suspend the clavicle and scapula from the back of the head, neck, and chest. Muscles E-G are the three shoulder muscles that originate from the anterior rib cage and insert onto the clavicle or scapula. Muscles H-K are the four rotator cuff muscles. Muscles L-O are the most powerful movers of the arm at the shoulder joint. Refer to this list in answering questions 11-19.

A. Trapezius
B. Levator scapulae
C. Rhomboid major
D. Rhomboid minor

E. Serratus anterior
F. Pectoralis minor
G. Subclavius

H. Supraspinatus
I. Infraspinatus
J. Subscapularis
K. Teres minor

L. Deltoid
M. Pectoralis major
N. Teres major
O. Latissimus dorsi

11. Which shoulder muscle is the prime mover for shrugging the shoulder? _____

12. Which shoulder muscle is the prime mover for protraction of the shoulder? _____

13. Which shoulder muscle is the prime mover for retraction of the shoulder? _____

14. Which shoulder muscle, starting from the anatomical position, initiates abduction of the arm at the shoulder joint? _____

15. Following 10° abduction of the arm from the anatomical position, which shoulder muscle becomes the sole prime mover for further arm abduction at the shoulder joint? _____

16. Abduction of the arm involves two coordinated movements: abduction of the arm at the shoulder joint and lateral rotation of the scapula. The answers to questions 14 and 15 have identified the two shoulder muscles responsible for abduction of the arm at the shoulder joint. Which two shoulder muscles are the prime movers for lateral rotation of the scapula? _____

17. Which three muscles are the most powerful internal rotators of the arm at the shoulder joint?

_____

18. Which two muscles are the most powerful external rotators of the arm at the shoulder joint?

_____

19. Which three muscles are the most powerful adductors of the arm at the shoulder joint?

_____

20. The shoulder joint is stabilized by both the rotator cuff muscles and ligaments in the fibrous capsule of the shoulder joint. Explain why the rotator cuff muscles stabilize the shoulder joint to a greater extent than the ligaments in the shoulder joint's fibrous capsule.

21. The rotator cuff muscles are so named because (a) each muscle can rotate the arm at the shoulder joint in a particular direction and (b) their insertion tendons collectively form a tendinous cuff around the head of the humerus. Explain why the insertion tendon of supraspinatus is more frequently inflamed or torn than the insertion tendons of the other three rotator cuff muscles.

Skin of the Shoulder

22. A dermatome is a strip of skin which receives most of its sensory innervation from a single spinal nerve. The skin that covers the upper surface of the shoulder includes parts of two dermatomes. These two dermatomes are the dermatomes of _____.

_____ C1
_____ C2
_____ C3
_____ C4
_____ C5
_____ C6
_____ C7
_____ C8
_____ T1
_____ T2
_____ T3
_____ T4
_____ T5

END OF QUESTIONS IN PART A OF THE CHAPTER ON THE SHOULDER

SHOULDER – Part B: Questions and Answers

Dissection of the shoulder in gross lab focuses on identification of the 15 shoulder muscles and some of the nerves that innervate these muscles. Three topics dominate the shoulder anatomy most frequently applied in clinical practice: (1) the nature of the three most common skeletal injuries of the shoulder, (2) the mechanism by which rotator cuff tendons commonly become inflamed or torn, and (3) knowledge of the prime movers of shoulder and arm movements.

Bones & Joints of the Shoulder

The clavicle, scapula, and the proximal, or upper, end of the humerus form the skeletal framework of the shoulder. The clavicle, scapula, and humerus form three synovial joints in the shoulder: the sternoclavicular, acromioclavicular, and shoulder joints. The three most common skeletal injuries of the shoulder are clavicular fractures, shoulder separations, and shoulder dislocations.

1. The clavicle is palpable along _____ in the upper anterior border of the shoulder.
   _____ only the medial third of its length
   _____ only the middle third of its length
   _____ only the lateral third of its length
   _____ only the medial and middle thirds of its length
   _____ only the middle and lateral thirds of its length
   __x__ its entire length

2. The five most prominent parts of the scapula are its acromion, coracoid process, glenoid cavity, spine, and inferior angle. Which of these parts are palpable during a physical exam of the shoulder?
   __x__ Acromion
   __x__ Coracoid process
   _____ Glenoid cavity
   __x__ Spine of the scapula
   __x__ Inferior angle of the scapula

3. Explain why the clavicle is the most commonly fractured bone of the body.

   The high incidence of clavicular fractures stems from the fact that the clavicle is always involved in transmitting forces and stresses from the upper limb to the body trunk when a person uses the upper limb to brace the body during accidental collisions and falls. The clavicle is always involved in transmitting forces and stresses between the upper limb and the body trunk because it is the only bone of the upper limb attached to the rib cage. Such forces and stresses, if excessive, put the clavicle at risk of fracture because the cross-sectional thickness of the shaft of the clavicle is relatively small.

4. Describe the nature of the injuries and the radiographic criteria of grade I, grade II, and grade III shoulder separations (shoulder separations are graded by comparing AP shoulder radiographs of the patient's injured and non-injured shoulders).

   Shoulder separations are injuries of the acromioclavicular (AC) joint. A shoulder separation does not involve any injury to the shoulder joint. Shoulder separations are commonly caused by a downward blow on the acromion. Such blows stress the two main structures that suspend the scapula from the clavicle: the fibrous capsule of the AC joint and the coracoclavicular ligament.

Shoulder separations are graded I to III according to the widths of two spaces in an AP radiograph of a patient's shoulder: the acromioclavicular space (which is the space between the lateral end of the clavicle and the acromion) and the coracoclavicular space (which is the space between the coracoid process and the clavicle above) (Fig. 1-1A). The acromioclavicular space represents the joint space of the AC joint. The coracoclavicular space marks the location of the coracoclavicular ligament.

A grade I shoulder separation is a simple sprain of the fibrous capsule of the AC joint. Neither the fibrous capsule nor the coracoclavicular ligament is significantly damaged. An AP radiograph of a patient's grade I shoulder separation thus shows acromioclavicular and coracoclavicular spaces of normal width (Fig. 1-1A).

A grade II shoulder separation is a subluxation (that is, a partial dislocation) of the AC joint. Subluxation occurs when significant damage to the joint's ligamentous supports is limited to its fibrous capsule. An AP radiograph of a patient's grade II shoulder separation shows a coracoclavicular space of normal width but an acromioclavicular space that is at least 50% wider than that measured in an AP radiograph of the patient's contralateral, uninjured shoulder (Fig 1-1B).

A grade III shoulder separation is a dislocation of the AC joint. Dislocation occurs when both the joint's fibrous capsule and the coracoclavicular ligament are significantly disrupted. An AP radiograph of a patient's grade III shoulder separation shows the acromioclavicular and coracoclavicular spaces both to be at least 50% wider than the corresponding spaces of the patient's contralateral, uninjured shoulder (Fig 1-1C).

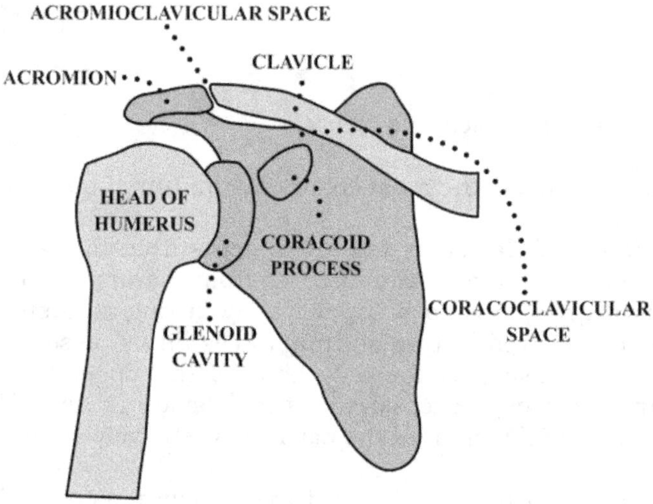

Fig. 1-1A: Outlines of the lateral half of the clavicle, scapula, and proximal humerus in an AP shoulder radiograph of a normal shoulder or a shoulder with a Grade I shoulder separation

ACROMIOCLAVICULAR SPACE

ACROMION

CLAVICLE

HEAD OF HUMERUS

CORACOID PROCESS

GLENOID CAVITY

CORACOCLAVICULAR SPACE

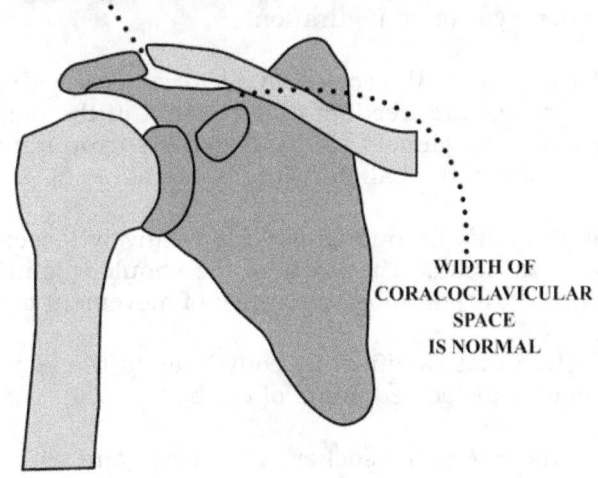

WIDTH OF ACROMIOCLAVICULAR SPACE INCREASED BY AT LEAST 50%

WIDTH OF CORACOCLAVICULAR SPACE IS NORMAL

Fig. 1-1B: Outlines of the lateral half of the clavicle, scapula, and proximal humerus in an AP shoulder radiograph of a shoulder with a Grade II shoulder separation; the acromioclavicular joint is subluxed

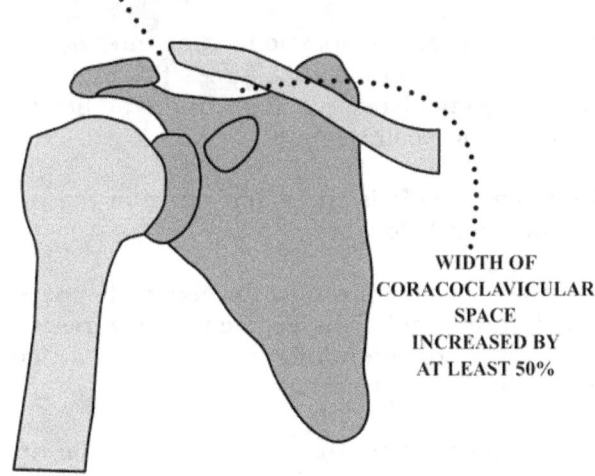

WIDTH OF ACROMIOCLAVICULAR SPACE SIGNIFICANTLY INCREASED

WIDTH OF CORACOCLAVICULAR SPACE INCREASED BY AT LEAST 50%

Fig. 1-1C: Outlines of the lateral half of the clavicle, scapula, and proximal humerus in an AP shoulder radiograph of a shoulder with a Grade III shoulder separation; the acromioclavicular joint is dislocated

5. The shoulder joint provides the greatest range and freedom of movement of any joint in the body. List the three anatomical factors that significantly enhance the range and freedom of movement of the arm in the shoulder joint.

(1) The shoulder joint has a ball-and-socket configuration. The synovial joints in the body with a ball-and-socket configuration provide greater range and freedom of movement than joints of any other type or configuration.

(2) The smaller the size of the socket in a syovial joint with a ball-and-socket configuration, the greater the range and freedom of movement of the ball. The size of the socket part of the shoulder joint (the glenoid cavity and the surrounding glenoid labrum) is very small compared to the size of the ball part (the humeral head).

(3) The shoulder joint's fibrous capsule is relatively loose; its surface area is about twice that of the humeral head. The laxity of the shoulder joint's fibrous capsule (especially its inferior part) markedly enhances the range of movement of the humeral head.

6. List the two factors that significantly contribute to the fact that the shoulder joint is one of the most commonly dislocated joints of the body.

(1) The smaller the size of the socket in a syovial joint with a ball-and-socket configuration, the lesser the contribution of the socket to the stability of the joint. The glenoid cavity is so small that it does not grip and help hold in place the humeral head.

(2) There are no ligaments or muscles that support the inferior, excessively lax part of the shoulder joint's fibrous capsule. This lack of inferior support tissues contributes to the relative ease with which the humeral head can be displaced anteroinferiorly from the glenoid cavity.

7. There are two types of shoulder dislocations: anterior and posterior. What is the approximate percentage of shoulder dislocations that are anterior?

More than 90% of shoulder dislocations are anterior shoulder dislocations. Anterior dislocation is produced by trauma that forces the humeral head anteroinferiorly from its articulation with the glenoid cavity. The humeral head commonly comes to rest in a subglenoid or subcoracoid position.

8. Explain why the axillary nerve is the major nerve of the upper limb most at risk of injury from an anterior shoulder dislocation.

As the axillary nerve extends through the axilla, it passes directly beneath the fibrous capsule of the shoulder joint. This segment of the nerve is stretched downward as the humeral head is displaced anteroinferiorly from its articulation with the glenoid cavity during an anterior shoulder dislocation.

9. Explain why the axillary nerve is the major nerve of the upper limb most at risk of injury from a fracture of the surgical neck of the humerus.

The segment of the axillary nerve which passes directly beneath the fibrous capsule of the shoulder joint in the axilla lies directly medial to the surgical neck of the humerus.

10. Abducting the arm 180° from the anatomical position involves coordinated movements at the sternoclavicular and shoulder joints. Describe the movements that occur at each joint during (a) abduction from 0 to 90° and (b) abduction from 90 to 180°.

During arm abduction from 0 to 90°, the arm is rotated upward at the shoulder joint as the shoulder girdle (that is, the clavicle and scapula together) is rotated upward at the sternoclavicular joint. The upward rotation of the arm at the shoulder joint occurs by the humeral head rolling upward on the glenoid cavity at the same time as the humeral head slides downward on the glenoid cavity. For every 2° abduction of the arm at the shoulder joint, there occurs 1° upward rotation of the shoulder girdle at the sternoclavicular joint. The upward rotation of the shoulder girdle at the sternoclavicular joint swings the inferior angle of the scapula upward and laterally; such movement by the inferior angle of the scapula is called lateral rotation of the scapula. It is important to recognize that the scapula cannot be rotated upward at the acromioclavicular joint because the joint does not provide for such a movement. Consequently, lateral rotation of the scapula during arm abduction from 0 to 90° occurs exclusively through upward rotation of the clavicle and scapula together at the sternoclavicular joint.

During arm abduction from 90 to 180°, the arm continues to be rotated upward at the shoulder joint. However, starting at about 90° abduction, the costoclavicular ligament (which is the ligament that stabilizes the sternoclavicular joint) becomes taut and prevents any further upward rotation of the clavicle at the sternoclavicular joint. At this juncture, lateral rotation of the scapula continues to occur as a result of posterior rotation of the clavicle at the sternoclavicular joint. Posterior rotation of the clavicle at the sternoclavicular joint is a movement in which the clavicle rotates 90° around its longitudinal axis in such a fashion that the undersurface of the clavicle is flipped upward and forward to become the anterior surface. This upward and forward flipping of the clavicle's undersurface swings the inferior angle of the scapula upward and laterally another 30°. In other words, the upward and forward flipping of the clavicle's under-surface produces another 30° lateral rotation of the scapula. 180° arm abduction is thus achieved by the combination of 120° upward rotation of the arm at the shoulder joint with 60° lateral rotation of the scapula.

Muscles of the Shoulder

The 15 shoulder muscles are listed here from A to O. Muscles A-D are the four shoulder muscles that suspend the clavicle and scapula from the back of the head, neck, and chest. Muscles E-G are the three shoulder muscles that originate from the anterior rib cage and insert onto the clavicle or scapula. Muscles H-K are the four rotator cuff muscles. Muscles L-O are the most powerful movers of the arm at the shoulder joint. Refer to this list in 0answering questions 11-19.

A. Trapezius	E. Serratus anterior	H. Supraspinatus	L. Deltoid
B. Levator scapulae	F. Pectoralis minor	I. Infraspinatus	M. Pectoralis major
C. Rhomboid major	G. Subclavius	J. Subscapularis	N. Teres major
D. Rhomboid minor		K. Teres minor	O. Latissimus dorsi

11. Which shoulder muscle is the prime mover for shrugging the shoulder? __A__

12. Which shoulder muscle is the prime mover for protraction of the shoulder? __E__

13. Which shoulder muscle is the prime mover for retraction of the shoulder? ___A__

14. Which shoulder muscle, starting from the anatomical position, initiates abduction of the arm at the shoulder joint? __H__

15. Following 10° abduction of the arm from the anatomical position, which shoulder muscle becomes the sole prime mover for further arm abduction at the shoulder joint? __L__

16. Abduction of the arm involves two coordinated movements: abduction of the arm at the shoulder joint and lateral rotation of the scapula. The answers to questions 14 and 15 have identified the two shoulder muscles responsible for abduction of the arm at the shoulder joint. Which two shoulder muscles are the prime movers for lateral rotation of the scapula? __A_E__

17. Which three muscles are the most powerful internal rotators of the arm at the shoulder joint?
__M_N_O__

18. Which two muscles are the most powerful external rotators of the arm at the shoulder joint?
__I_K__

19. Which three muscles are the most powerful adductors of the arm at the shoulder joint?
__M_N_O__

20. The shoulder joint is stabilized by both the rotator cuff muscles and ligaments in the fibrous capsule of the shoulder joint. Explain why the rotator cuff muscles stabilize the shoulder joint to a greater extent than the ligaments in the shoulder joint's fibrous capsule.

In most instances, the ligaments that bind together the bones of a joint stabilize the joint only at the limits of the movements at the joint. Muscles, however, can serve as extensile ligaments that resist disruptive forces throughout the entire range of movements at the joint. It is this capacity of muscles to exert cohesive forces across a joint throughout its entire range of movements that accounts for the fact that the tone of the muscles acting across a joint is the most important factor in stabilizing the joint.

16

21. The rotator cuff muscles are so named because (a) each muscle can rotate the arm at the shoulder joint in a particular direction and (b) their insertion tendons collectively form a tendinous cuff around the head of the humerus. Explain why the insertion tendon of supraspinatus is more frequently inflamed or torn than the insertion tendons of the other three rotator cuff muscles.

The region immediately overlying the shoulder joint's fibrous capsule is a relatively narrow space called the subacromial space. It is bordered superiorly by both the acromion and the coracoacromial ligament and inferiorly by the shoulder joint's fibrous capsule. The coracoacromial ligament is a ligament extending between two parts of the scapula: its coracoid process and the acromion. Two tissues lie within the subacromial space: supraspinatus' insertion tendon and the subacromial bursa (the subacromial bursa is the largest bursa of the body). Supraspinatus' insertion tendon and the subacromial bursa move within the subacromial space whenever the arm is abducted or adducted at the shoulder joint. Supraspinatus' insertion tendon and the subacromial bursa may become impinged (that is, pressed upon) when the arm is simultaneously abducted, flexed, and internally rotated at the shoulder joint, as this is the arm position at which the subacromial space becomes narrowest. Continued, excessive impingement commonly results in inflammation or tearing of supraspinatus' insertion tendon and/or inflammation of the subacromial bursa. These injuries are collectively referred to as the subacromial impingement syndrome. The insertion tendons of the other three rotator cuff muscles are at less risk of impingement in the subacromial space than supraspinatus' insertion tendon. This lesser risk accounts for supraspinatus' insertion tendon being more frequently inflamed or torn than the insertion tendons of the other three rotator cuff muscles.

Skin of the Shoulder

22. A dermatome is a strip of skin which receives most of its sensory innervation from a single spinal nerve. The skin that covers the upper surface of the shoulder includes parts of two dermatomes. These two dermatomes are the dermatomes of _____.
　_____ C1
　_____ C2
　_x_ C3
　_x_ C4
　_____ C5
　_____ C6
　_____ C7
　_____ C8
　_____ T1
　_____ T2
　_____ T3
　_____ T4
　_____ T5

18

AXILLA – Part A: Questions

The axilla, or armpit, is the upper limb region that extends from the shoulder to the arm. Dissection of the axilla in gross lab focuses on identification of the parts of the brachial plexus and the branches of the axillary artery. Two topics dominate the axillary anatomy most frequently applied in clinical practice: (1) knowledge of the structure of the brachial plexus and (2) the location and drainage areas of the lymph node groups clustered in and around the axilla.

The axilla is bordered anteriorly by a tissue fold called the anterior axillary fold and posteriorly by a tissue fold called the posterior axillary fold. Pectoralis major comprises the muscle mass in the anterior axillary fold; latissimus dorsi and teres major comprise the muscle mass in the posterior axillary fold. The axillary artery, axillary vein, and brachial plexus extend through the axilla. The axilla and immediate surrounding regions contain almost all the lymph nodes that drain lymph from the upper limb. The brachial plexus is the nerve network that gives rise to almost all the nerves that provide motor and sensory innervation for the upper limb. The brachial plexus consists of 5 parts: its roots, trunks, divisions, cords, and nerves.

Brachial Plexus

1. The anterior rami of 5 spinal nerves form the roots of the brachial plexus. The roots of the brachial plexus are located entirely in the neck. The roots of the brachial plexus are formed from the anterior rami of which 5 spinal nerves?

    _____ C1   _____ C5   _____ T1
    _____ C2   _____ C6   _____ T2
    _____ C3   _____ C7   _____ T3
    _____ C4   _____ C8   _____ T4

2. The roots of the brachial plexus give rise to three trunks. Complete the following sentences:
    (a) The _____ roots unite to form the upper trunk.
    (b) The middle trunk is a continuation of the _____ root.
    (c) The _____ roots unite to form the lower trunk.

Each trunk divides into an anterior division and a posterior division. The divisions of the brachial plexus unite to form the three cords of the brachial plexus. The cords are named for their relationship to the axillary artery. The lateral, posterior, and medial cords lie, respectively, lateral, posterior, and medial to the axillary artery. Most of the nerves that arise from the brachial plexus arise from the cords. In particular, the 5 major nerves of the upper limb all arise from the cords.

Complete the following sentences by identifying the cord or cords which give rise to each major nerve of the upper limb:

3. The axillary nerve arises from the _____ cord.

4. The median nerve arises from both the _____ cords.

5. The musculocutaneous nerve arises from the _____ cord.

6. The radial nerve arises from the _____ cord.

7. The ulnar nerve arises from the _____ cord.

Identify the nerve or nerves which innervate each shoulder muscle:

8. Trapezius: _____

9. Levator scapulae: _____

10. Rhomboid major: _____

11. Rhomboid minor: _____

12. Serratus anterior: _____

13. Pectoralis minor: _____

14. Subclavius: _____

15. Supraspinatus: _____

16. Infraspinatus: _____

17. Subscapularis _____

18. Teres minor: _____

19. Deltoid: _____

20. Pectoralis major: _____

21. Teres major: _____

22. Latissimus dorsi: _____

Shoulder Muscles Which Receive All Their Innervation from Only C5 and C6 Nerve Fibers

23. Near the end of your study of the forearm and wrist in this workbook, we will examine upper brachial plexus injuries. At that time, you will have to recall which shoulder and arm muscles receive all or most of their innervation from only or mainly C5 and C6 nerve fibers. Identify the 4 shoulder muscles that receive ALL of their innervation from ONLY C5 and C6 nerve fibers: _____

A. Trapezius	E. Serratus anterior	H. Supraspinatus	L. Deltoid
B. Levator scapulae	F. Pectoralis minor	I. Infraspinatus	M. Pectoralis major
C. Rhomboid major	G. Subclavius	J. Subscapularis	N. Teres major
D. Rhomboid minor		K. Teres minor	O. Latissimus dorsi

24. Explain why it can be said that C5 controls abduction of the arm at the shoulder joint.

Axillary Lymph Nodes

25. There are six groups of lymph nodes in and around the axilla. Describe the tissues from which each lymph node group drains lymph.

26. Describe the general physical features of normal superficial lymph nodes.

27. Describe the general physical features of superficial lymph nodes reacting to an infection.

28. Describe the general physical features of superficial lymph nodes invaded by malignant cells.

29. In addition to the anterior group of axillary lymph nodes and the internal thoracic lymph nodes, which other lymph node groups filter some of the lymph drained from the breast?

END OF QUESTIONS IN PART A OF THE CHAPTER ON THE AXILLA

AXILLA – Part B: Questions and Answers

The axilla, or armpit, is the upper limb region that extends from the shoulder to the arm. Dissection of the axilla in gross lab focuses on identification of the parts of the brachial plexus and the branches of the axillary artery. Two topics dominate the axillary anatomy most frequently applied in clinical practice: (1) knowledge of the structure of the brachial plexus and (2) the location and drainage areas of the lymph node groups clustered in and around the axilla.

The axilla is bordered anteriorly by a tissue fold called the anterior axillary fold and posteriorly by a tissue fold called the posterior axillary fold. Pectoralis major comprises the muscle mass in the anterior axillary fold; latissimus dorsi and teres major comprise the muscle mass in the posterior axillary fold. The axillary artery, axillary vein, and brachial plexus extend through the axilla. The axilla and immediate surrounding regions contain almost all the lymph nodes that drain lymph from the upper limb. The brachial plexus is the nerve network that gives rise to almost all the nerves that provide motor and sensory innervation for the upper limb. The brachial plexus consists of 5 parts: its roots, trunks, divisions, cords, and nerves.

Brachial Plexus

1. The anterior rami of 5 spinal nerves form the roots of the brachial plexus. The roots of the brachial plexus are located entirely in the neck. The roots of the brachial plexus are formed from the anterior rami of which 5 spinal nerves?

_____ C1	__x__ C5	__x__ T1
_____ C2	__x__ C6	_____ T2
_____ C3	__x__ C7	_____ T3
_____ C4	__x__ C8	_____ T4

2. The roots of the brachial plexus give rise to three trunks. Complete the following sentences:
   (a) The __C5 and C6__ roots unite to form the upper trunk.
   (b) The middle trunk is a continuation of the __C7__ root.
   (c) The __C8 and T1__ roots unite to form the lower trunk.

Each trunk divides into an anterior division and a posterior division. The divisions of the brachial plexus unite to form the three cords of the brachial plexus. The cords are named for their relationship to the axillary artery. The lateral, posterior, and medial cords lie, respectively, lateral, posterior, and medial to the axillary artery. Most of the nerves that arise from the brachial plexus arise from the cords. In particular, the 5 major nerves of the upper limb all arise from the cords.

Complete the following sentences by identifying the cord or cords which give rise to each major nerve of the upper limb:

3. The axillary nerve arises from the __posterior__ cord.

4. The median nerve arises from both the __medial__ and __lateral__ cords.

5. The musculocutaneous nerve arises from the __lateral__ cord.

6. The radial nerve arises from the __posterior__ cord.

7. The ulnar nerve arises from the __medial__ cord.

23

Identify the nerve or nerves which innervate each shoulder muscle:

8. Trapezius:__accessory nerve__ (Trapezius is the only upper limb muscle not innervated by a nerve that arises from the brachial plexus.)

9. Levator scapulae: __dorsal scapular nerve (and C3 and C4 nerve fibers)__

10. Rhomboid major: __dorsal scapular nerve__

11. Rhomboid minor: __dorsal scapular nerve__

12. Serratus anterior: __long thoracic nerve__

13. Pectoralis minor: __medial pectoral nerve__

14. Subclavius: __nerve to subclavius__

15. Supraspinatus: __suprascapular nerve__

16. Infraspinatus: __suprascapular nerve__

17. Subscapularis __upper and lower subscapular nerves__

18. Teres minor: __axillary nerve__

19. Deltoid: __axillary nerve__

20. Pectoralis major: __medial and lateral pectoral nerves__

21. Teres major: __lower subscapular nerve__

22. Latissimus dorsi: __thoracodorsal nerve (also called the middle subscapular nerve)

Shoulder Muscles Which Receive All Their Innervation from Only C5 and C6 Nerve Fibers

23. Near the end of your study of the forearm and wrist in this workbook, we will examine upper brachial plexus injuries. At that time, you will have to recall which shoulder and arm muscles receive all or most of their innervation from only or mainly C5 and C6 nerve fibers. Identify the 4 shoulder muscles that receive ALL of their innervation from ONLY C5 and C6 nerve fibers: __H__I__K__L__

A. Trapezius	E. Serratus anterior	H. Supraspinatus	L. Deltoid
B. Levator scapulae	F. Pectoralis minor	I. Infraspinatus	M. Pectoralis major
C. Rhomboid major	G. Subclavius	J. Subscapularis	N. Teres major
D. Rhomboid minor		K. Teres minor	O. Latissimus dorsi

24. Explain why it can be said that C5 controls abduction of the arm at the shoulder joint.

Abduction of the arm at the shoulder joint is controlled mainly by C5 nerve fibers because supraspinatus and deltoid, which are the only muscles involved in abduction of the arm at the shoulder joint, each receive more innervation from C5 than C6 nerve fibers. The answers to questions 15 and 16 in the chapter on the shoulder identified supraspinatus and deltoid as the only muscles involved in abduction of the arm at the shoulder joint.

Knowledge that C5 controls abduction of the arm at the shoulder joint is important in the examination of patients who present with muscle weakness, pain, and/or diminshed sensation in the upper limb. One of the relatively common causes of such symptoms, especially among adults 40 years of age and older, is impingement in the neck of one of the 5 spinal nerves that form the roots of the brachial plexus. The impingement commonly occurs in the vicinity of the intervertebral foramen through which the spinal nerve exits the spine. If C5 is impinged and the impingement has caused motor nerve fibers in C5 to be injured, supraspinatus and deltoid may be weakened. Such weakness can be assessed by testing the strength of arm abduction against resistance in both the affected and unaffected upper limbs.

Axillary Lymph Nodes

25. There are six groups of lymph nodes in and around the axilla. Describe the tissues from which each lymph node group drains lymph:

Anterior group (which can be palpated against the posterior surface of the anterior axillary fold): Superficial tissues of the anterior trunk of the body, down to the level of the umbilicus (includes the tissues in the lateral half of the breast)

Posterior group (which can be palpated against the anterior surface of the posterior axillary fold): Superficial tissues of the posterior trunk of the body, down to the level of the iliac crest

Lateral group (which can be palpated against the medial surface of the uppermost part of the shaft of the humerus): Deep tissues of the hand, forearm, and arm and the superficial tissues of the medial side of the upper limb

Central group (which can be palpated against the rib cage in the center of the axilla): The efferent lymphatics that conduct lymph from the anterior, posterior, and lateral groups

Infraclavicular group (which can be palpated below the middle third of the clavicle): Superficial tissues of the lateral side of the hand, forearm, and arm

Apical group (which cannot be palpated and lie immediately lateral to the lateral border of the 1st rib): The efferent lymphatics that conduct lymph from the other 5 axillary groups

Breast

When you examine the breasts of a patient, you should always recall the anatomical associations of the lateral half of the breast with the axilla. There are 3 anatomical associations: (1) A branch of the axillary artery (the lateral thoracic artery) is the chief source of blood supply to the lateral half of the breast. (2) Most of the venous blood drained from the lateral half of the breast flows into the axillary vein. (3) Most of the lymph drained from the lateral half of the breast is filtered first by the anterior group of axillary lymph nodes and then secondarily by the central group.

Branches of the internal thoracic artery are the chief source of blood supply to the medial half of the breast. Most of the blood drained from the medial half of the breast flows into the internal thoracic veins. Most of the lymph drained from the medial half of the breast is filtered first by the internal thoracic nodes (also called the parasternal or internal mammary nodes). The internal thoracic nodes are not palpable.

26. Describe the general physical features of normal superficial lymph nodes.

   Normal superficial lymph nodes are frequently difficult to palpate because they are small (one cm or less in length), non-tender, relatively mobile, and soft.

27. Describe the general physical features of superficial lymph nodes reacting to an infection.

   Lymph nodes that are mounting an immunological response to infectious agents are enlarged but regularly shaped, tender, relatively mobile, and firm.

28. Describe the general physical features of superficial lymph nodes invaded by malignant cells.

   Lymph nodes involved in carcinamatous metastasis are non-tender but become hard, fixed, and irregularly shaped.

29. In addition to the anterior group of axillary lymph nodes and the internal thoracic lymph nodes, which other lymph node groups filter some of the lymph drained from the breast?

   In the neck, there are lymph nodes called deep cervical nodes that are aligned alongside the internal jugular vein. The lowest deep cervical nodes may be referred to as supraclavicular lymph nodes because they lie immediately superior to the medial third of the clavicle. Supraclavicular lymph nodes filter some of the lymph drained from the superior region of the breast. Some lymph drained from a breast may also be filtered by the internal thoracic, supraclavicular, and even axillary lymph nodes on the contralateral side of the body.

Dissection of the arm and elbow in gross lab focuses on identification of the 4 arm muscles; the median, musculocutaneous, radial, and ulnar nerves; the brachial and deep brachial arteries; and the basilic and cephalic veins. Five topics dominate the arm and elbow anatomy most frequently applied in clinical practice: (1) the use of the brachial artery in the mid arm region as the best site to measure blood pressure, (2) the use of the subcutaneous veins of the cubital fossa for drawing venous blood from a patient, (3) the innervation and major actions of the muscles of the arm, (4) the spinal cord levels at which spinal cord reflexes are assessed by the biceps brachii and triceps deep tendon reflexes, and (5) knowledge of the major upper limb nerves at risk of injury from midhumeral shaft and medial epicondyle fractures.

Bones & Joints Associated with the Actions of Arm Muscles

1. Complete the following statements concerning bony landmarks around the elbow:

    The prominent bony bumps that can be palpated on the medial and lateral sides of the elbow are the _____.

    The prominent bony bump that can be palpated at the back of the elbow is the _____.

    It is __(True or False)__ that the head of the radius can be palpated on the lateral side of the elbow.

2. In examining a lateral radiograph of the elbow, a radiologist observes two abnormal signs called the sail and posterior fat pad signs. What is the anatomical basis of these signs, and what is their significance?

3. A pulled, or nursemaid's, elbow is a common elbow injury among young children. Describe the common history of a pulled elbow and the nature of the injury. What are the radiographic signs of a pulled elbow? What is the procedure by which a pulled elbow can be reduced in the office setting?

4. Of the 5 major nerves of the upper limb (the axillary, median, musculocutaneous, radial, and ulnar nerves), only 4 extend from the shoulder down the arm. The axillary nerve ends in the axilla, where it innervates just two muscles (deltoid and teres minor) with only C5 and C6 nerve fibers. Which of the 4 remaining major nerves of the upper limb is most at risk of injury from a fracture of the midshaft of the humerus? Explain the anatomical basis of the risk.

5. Which of the 4 major nerves of the upper limb that extend distally through the arm is most at risk of injury from a fracture of the medial epicondyle of the humerus? Explain the anatomical basis of the risk.

Muscles of the Arm

The 4 muscles of the arm are listed here from A to D:
    A. Coracobrachialis   B. Biceps brachii   C. Brachialis   D. Triceps

Identify the nerve which innervates each arm muscle:

6. Coracobrachialis: _____

7. Biceps brachii: _____

8. Brachialis: _____

9. Triceps: _____

Biceps brachii, brachialis, and triceps are prime movers of the forearm. Identify the action or actions for which each of these arm muscles is a prime mover:

10. Biceps brachii is one of the two prime movers of:
    _____ Extension of the forearm at the elbow joint
    _____ Flexion of the forearm at the elbow joint
    _____ Pronation of the forearm at the proximal and distal radioulnar joints
    _____ Supination of the forearm at the proximal and distal radioulnar joints

11. Brachialis is one of the two prime movers of:
    _____ Extension of the forearm at the elbow joint
    _____ Flexion of the forearm at the elbow joint
    _____ Pronation of the forearm at the proximal and distal radioulnar joints
    _____ Supination of the forearm at the proximal and distal radioulnar joints

12. Triceps is the sole prime mover of:
    _____ Extension of the forearm at the elbow joint
    _____ Flexion of the forearm at the elbow joint
    _____ Pronation of the forearm at the proximal and distal radioulnar joints
    _____ Supination of the forearm at the proximal and distal radioulnar joints

13. The biceps brachii tendon reflex test assesses spinal cord reflex activity at the two spinal cord levels which provide most of the motor and sensory nerve fibers to biceps brachii. Identify these two spinal cord levels.
    _____ C5
    _____ C6
    _____ C7
    _____ C8
    _____ T1

14. The triceps tendon reflex test assesses spinal cord reflex activity at the two spinal cord levels which provide most of the motor and sensory nerve fibers to triceps. Identify these two spinal cord levels.
    _____ C5
    _____ C6
    _____ C7
    _____ C8
    _____ T1

15. Explain why it can be said that C6 controls flexion of the forearm at the elbow joint.

16. Explain why it can be said that C7 controls extension of the forearm at the elbow joint.

Skin of the Arm

17. Most of the skin on the lateral side of the arm is part of a dermatome; in other words, one spinal nerve provides most of the cutaneous sensory nerve fibers for the skin on the lateral side of the arm. Identify the spinal nerve.

____ C1	____ T1
____ C2	____ T2
____ C3	____ T3
____ C4	____ T4
____ C5	____ T5
____ C6	
____ C7	
____ C8	

18. The skin on the medial side of the arm includes parts of two dermatomes. These two dermatomes are the dermatomes of ____.

____ C1	____ T1
____ C2	____ T2
____ C3	____ T3
____ C4	____ T4
____ C5	____ T5
____ C6	
____ C7	
____ C8	

Brachial Artery, Superficial Cubital Veins, and Cubital Lymph Nodes

19. Explain why the mid arm region is the best site to measure blood pressure.

20. Which major nerve of the upper limb extends alongside the brachial artery as it extends distally through the arm and cubital fossa? _____

21. What is commonly the largest subcutaneous (superficial) vein of the cubital fossa?

_____

22. The most distal lymph nodes of the upper limb are superficial lymph nodes located near the elbow. Describe their location and the tissues from which they drain lymph.

END OF QUESTIONS IN PART A OF THE CHAPTER ON THE ARM AND ELBOW

ARM AND ELBOW – Part B: Questions and Answers

Dissection of the arm and elbow in gross lab focuses on identification of the 4 arm muscles; the median, musculocutaneous, radial, and ulnar nerves; the brachial and deep brachial arteries; and the basilic and cephalic veins. Five topics dominate the arm and elbow anatomy most frequently applied in clinical practice: (1) the use of the brachial artery in the mid arm region as the best site to measure blood pressure, (2) the use of the subcutaneous veins of the cubital fossa for drawing venous blood from a patient, (3) the innervation and major actions of the muscles of the arm, (4) the spinal cord levels at which spinal cord reflexes are assessed by the biceps brachii and triceps deep tendon reflexes, and (5) knowledge of the major upper limb nerves at risk of injury from midhumeral shaft and medial epicondyle fractures.

Bones & Joints Associated with the Actions of Arm Muscles

The 4 muscles of the arm move the forearm at 3 synovial joints: the elbow, proximal radioulnar, and distal radioulnar joints. The elbow joint provides for flexion and extension of the forearm, and the proximal and distal radioulnar joints provide for supination and pronation of the forearm. In the elbow joint, the capitulum of the humerus articulates with the head of the radius in the lateral side of the joint, and the trochlea of the humerus articulates with the trochlear notch of the ulna in the medial side of the joint. In the proximal radioulnar joint, the head of the radius articulates with a small notch (called the radial notch) near the proximal end of the ulna. In the distal radioulnar joint, the distal end of the ulna (which is called the head of the ulna) articulates with a small notch (called the ulnar notch) at the distal end of the radius.

1. Complete the following statements concerning bony landmarks around the elbow:

   The prominent bony bumps that can be palpated on the medial and lateral sides of the elbow are the __medial and lateral epicondyles of the humerus__.

   The prominent bony bump that can be palpated at the back of the elbow is the __olecranon process of the ulna__.

   It is __true__ that the head of the radius can be palpated on the lateral side of the elbow.

2. In examining a lateral radiograph of the elbow, a radiologist observes two abnormal signs called the sail and posterior fat pad signs. What is the anatomical basis of these signs, and what is their significance?

   Near the lower end of the humerus, there are deep recesses in the anterior and posterior surfaces of the humerus immediately above the capitulum and trochlea; these recesses are called, respectively, the coronoid and olecranon fossae. In the elbow, pads of fat tissue called fat pads overlie the coronoid and olecranon fossae. The fat pad overlying the coronoid fossa is called the anterior fat pad, and the fat pad overlying the olecranon fossa is called the posterior fat pad. A lateral radiograph of a normal elbow joint shows the presence of the anterior but not the posterior fat pad (Fig. 3-1A). When the forearm is flexed 90° at the elbow joint, the superficial part of the anterior fat pad lies anterior to the distal end of the humerus. Fat tissue casts a darker shadow in radiographs than muscle tissue because fat tissue is more radiolucent (that is, it interferes less with X-rays passing through it) than muscle tissue. Therefore, in a lateral radiograph of a normal elbow, the superficial part of the anterior fat pad appears as a thin dark band sandwiched between the highly radiopaque, or whitish, bony rim of the humerus and the moderately radiopaque, or grayish, shadows of the overlying, anterior arm muscles (Fig. 3-1A). The posterior fat pad cannot be seen in a normal lateral radiograph of the elbow because the posterior fat pad lies deep to the bony rim of the humerus.

When injury or infection of the elbow joint produces an effusion (that is, an increase of fluid) within the synovial cavity of the joint, the swollen joint pushes the anterior fat pad forward and the posterior fat pad backward. Consequently, a lateral radiograph of an effused elbow will not only show an enlarged anterior fat pad but also the presence of the posterior fat pad (Fig. 3-1B). The displacement of the anterior fat pad increases the size of the dark area immediately in front of the lower anterior end of the humerus; this radiographic sign is called the sail sign because the dark area is shaped like a triangular sail (Fig. 3-1B). The displacement of the posterior fat pad produces a dark area behind the lower posterior end of the humerus; this radiographic sign is called the posterior fat pad sign (Fig. 3-1B).

The sail and posterior fat pad signs indicate effusion of the elbow joint. In patients who have sustained an injury to the elbow, the signs typically indicate the occurrence of an intra-articular fracture (that is, a fracture involving a bony surface within the elbow joint's synovial cavity). In young children, the most common fracture is a supracondylar fracture (that is, a fracture at the distal end of the humerus immediately above the capitulum and trochlea). In adults, the most common fracture is a radial head fracture.

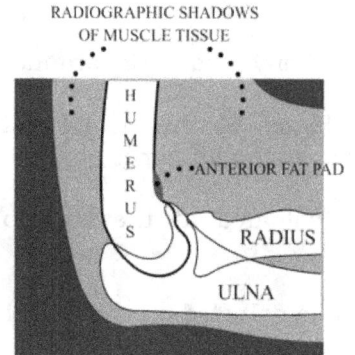 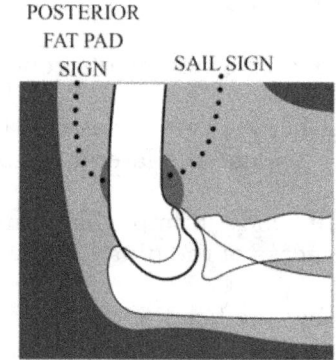

Fig. 3-1A: Lateral radiograph of normal elbow showing anterior fat pad   Fig. 3-1B: Lateral radiograph of effused elbow showing sail and posterior fat pad signs

3. A pulled, or nursemaid's, elbow is a common elbow injury among young children. Describe the common history of a pulled elbow and the nature of the injury. What are the radiographic signs of a pulled elbow? What is the procedure by which a pulled elbow can be reduced in the office setting?

In the proximal radioulnar joint, a C-shaped ligament called the annular ligament partially surrounds the head of the radius. The annular ligament stabilizes the articulation of the head of the radius with the radial notch of the ulna in the proximal radioulnar joint. When the forearm is pronated or supinated at the proximal and distal radioulnar joints, the head of the radius rotates within its cufflike annular ligament at the proximal radioulnar joint.

A pulled elbow generally results from a sudden distal pull on a young child's upper limb. The sudden distal pull results in the upper part of the annular ligament partially slipping over the upper surface of the head of the radius and becoming entrapped between the head of the radius and the capitulum of the humerus. When the injured child is examined, he/she typically presents with a flexed and partially pronated forearm supported closely to the trunk of the body; the child may complain of pain around the head of the radius. A lateral radiograph of the child's elbow will appear normal; the entrapment of part of the annular ligament between the head of the radius and the capitulum of the humerus does not displace the head of the radius. Reduction of a pulled elbow may be achieved in the office setting by stabilizing the child's elbow and palpating the head of the radius with one hand and then slowly, but firmly, supinating and extending the child's forearm with the other hand.

4. Of the 5 major nerves of the upper limb (the axillary, median, musculocutaneous, radial, and ulnar nerves), only 4 extend from the shoulder down the arm. The axillary nerve ends in the axilla, where it innervates just two muscles (deltoid and teres minor) with only C5 and C6 nerve fibers. Which of the 4 remaining major nerves of the upper limb is most at risk of injury from a fracture of the midshaft of the humerus? Explain the anatomical basis of the risk.

As the radial nerve extends distally through the arm, it enters the posterior muscular compartment by coursing alongside the spiral groove on the midshaft of the humerus. This relationship is the anatomical basis of the radial nerve being the major upper limb nerve most at risk of injury from a fracture of the midshaft of the humerus.

5. Which of the 4 major nerves of the upper limb that extend distally through the arm is most at risk of injury from a fracture of the medial epicondyle of the humerus? Explain the anatomical basis of the risk.

As the ulnar nerve extends from the arm into the forearm, it passes immediately posterior to the medial epicondyle of the humerus. This relationship is the anatomical basis of the ulnar nerve being the major upper limb nerve most at risk of injury from a fracture of the medial epicondyle.

Muscles of the Arm

The 4 muscles of the arm are listed here from A to D:
   A. Coracobrachialis   B. Biceps brachii   C. Brachialis   D. Triceps

Identify the nerve which innervates each arm muscle:

6. Coracobrachialis: __musculocutaneous nerve__

7. Biceps brachii: __musculocutaneous nerve__

8. Brachialis: __musculocutaneous nerve__

9. Triceps: __radial nerve__

Biceps brachii, brachialis, and triceps are prime movers of the forearm. Identify the action or actions for which each of these arm muscles is a prime mover:

10. Biceps brachii is one of the two prime movers of:
    _____ Extension of the forearm at the elbow joint
    __x__ Flexion of the forearm at the elbow joint
    _____ Pronation of the forearm at the proximal and distal radioulnar joints
    __x__ Supination of the forearm at the proximal and distal radioulnar joints

11. Brachialis is one of the two prime movers of:
    _____ Extension of the forearm at the elbow joint
    __x__ Flexion of the forearm at the elbow joint
    _____ Pronation of the forearm at the proximal and distal radioulnar joints
    _____ Supination of the forearm at the proximal and distal radioulnar joints
12. Triceps is the sole prime mover of:
    __x__ Extension of the forearm at the elbow joint
    _____ Flexion of the forearm at the elbow joint
    _____ Pronation of the forearm at the proximal and distal radioulnar joints
    _____ Supination of the forearm at the proximal and distal radioulnar joints

During a physical examination of a patient, it is common for a physician to use a reflex hammer to impart a gentle but firm and quick tap upon the insertion tendons of biceps brachii and triceps. The firm and quick tap suddenly stretches the muscle's insertion tendon, which, in turn, elicits a spinal cord reflex that produces a quick, short reflexive contraction of the muscle. These spinal cord reflexes are called deep tendon reflexes, and they occur at the spinal cord levels which provide motor and sensory nerve fibers to the muscle being tested. Deep tendon reflexes are commonly assessed in patients with neurological and/or muscular injuries or disorders.

13. The biceps brachii tendon reflex test assesses spinal cord reflex activity at the two spinal cord levels which provide most of the motor and sensory nerve fibers to biceps brachii. Identify these two spinal cord levels.
    __x__ C5
    __x__ C6
    _____ C7
    _____ C8
    _____ T1

14. The triceps tendon reflex test assesses spinal cord reflex activity at the two spinal cord levels which provide most of the motor and sensory nerve fibers to triceps. Identify these two spinal cord levels.
    _____ C5
    _____ C6
    __x__ C7
    __x__ C8
    _____ T1

15. Explain why it can be said that C6 controls flexion of the forearm at the elbow joint.

    Biceps brachii and brachialis are the two prime movers of flexion of the forearm at the elbow joint. Like biceps brachii, brachialis also receives most of its motor and sensory fibers from C5 and C6. Flexion of the forearm at the elbow joint is controlled mainly by C6 nerve fibers because biceps brachii and brachialis each receive more innervation from C6 than C5 nerve fibers.

34

16. Explain why it can be said that C7 controls extension of the forearm at the elbow joint.

Extension of the forearm at the elbow joint is controlled mainly by C7 nerve fibers because triceps, which is the sole prime mover of extension of the forearm at the elbow joint, receives more innervation from C7 than C8 nerve fibers.

Skin of the Arm

17. Most of the skin on the lateral side of the arm is part of a dermatome; in other words, one spinal nerve provides most of the cutaneous sensory nerve fibers for the skin on the lateral side of the arm. Identify the spinal nerve.

_____ C1
_____ C2
_____ C3
_____ C4
__x__ C5
_____ C6
_____ C7
_____ C8
_____ T1
_____ T2
_____ T3
_____ T4
_____ T5

18. The skin on the medial side of the arm includes parts of two dermatomes. These two dermatomes are the dermatomes of _____.

_____ C1
_____ C2
_____ C3
_____ C4
_____ C5
_____ C6
_____ C7
_____ C8
__x__ T1
__x__ T2
_____ T3
_____ T4
_____ T5

19. Explain why the mid arm region is the best site to measure blood pressure.

   There are two principal reasons: First, the brachial artery in the mid arm region lies approximately at the level of the heart. Two forces contribute to the blood pressure in every vessel of the body: the heart's pumping activity and the earth's gravitational field. The contribution of gravity is proportional to the vertical distance between the vessel in question and the heart. With the arm positioned by the chest (independently of whether the person is standing, sitting, or reclining), the brachial artery in the mid arm region is, for all practical purposes, lying at the level of the heart. Consequently, the blood pressure of the brachial artery in the mid arm region represents blood pressure generated almost exclusively by the heart's pumping activity alone.

   Second, the brachial artery is relatively deep and close to the humeral shaft as it extends distally through the mid arm region. The deep course of the brachial artery through the mid arm region makes it possible to stop its blood flow by wrapping an inflatable cuff around the mid arm and increasing the pressure in the cuff above the systolic pressure. Blood flow in the brachial artery stops because the muscle mass in the arm efficiently transmits the pressure in the cuff and sequeezes the brachial artery against the humeral shaft.

20. Which major nerve of the upper limb extends alongside the brachial artery as it extends distally through the arm and cubital fossa? __median nerve__

21. What is commonly the largest subcutaneous (superficial) vein of the cubital fossa? __median cubital vein__

22. The most distal lymph nodes of the upper limb are superficial lymph nodes located near the elbow. Describe their location and the tissues from which they drain lymph.

   The lymph nodes near the elbow are called the supratrochlear, or cubital, lymph nodes. They can be palpated along the medial side of the basilic vein immediately superior to the medial epicondyle of the humerus. They drain lymph from superficial tissues on the medial side of the forearm and hand.

FOREARM AND WRIST – Part A: Questions

Dissection of the forearm and wrist in gross lab focuses on identification of the anterior and posterior forearm muscles, the flexor and extensor retinacula, the median, anterior interosseous, deep radial, posterior interosseous, and ulnar nerves, and the radial and ulnar arteries. Five topics dominate the forearm anatomy most frequently applied in clinical practice: (1) the use of the radial and ulnar arteries at the wrists as the best sites to measure a patient's pulse, (2) knowledge of the courses by which the median and ulnar nerves extend from the forearm into the hand, (3) the use of the anatomical snuffbox for the examination of a fractured scaphoid, (4) the nature of a Colles' fracture, and (5) the innervation and major actions of the muscles of the forearm.

Bones & Joints Associated with the Actions of Forearm Muscles

The muscles of the forearm can move the forearm at the proximal and distal radioulnar joints, the hand at the wrist and midcarpal joints, and the thumb and fingers at their metacarpophalangeal (MCP) and interphalangeal (IP) joints.

In the following pages, as you answer questions related to the bones and joints of the hand, it is important to recognize that anatomically and medically the thumb is not a finger. In the hand, there are five digits, and they consist of the four fingers plus the thumb. Observe that each digit has two names. When you identify a digit of the hand in clinical notes, it is advisable to include both names of the digit:

Thumb-1st digit

Index finger-2nd digit

Middle finger-3rd digit

Ring finger-4th digit

Little finger-5th digit

1. The styloid process of the radius can be palpated at the distal end of the radius:
   True or False

2. When a person makes a fist, which bony surfaces underlie the knuckles?
   _____ Bases of the 2nd, 3rd, 4th and 5th metacarpals
   _____ Bases of the proximal phalanges of the fingers
   _____ Heads of the 2nd, 3rd, 4th, and 5th metacarpals
   _____ Heads of the proximal phalanges of the fingers

3. Which carpal is the most commonly fractured carpal?
   _____ Capitate   _____ Scaphoid
   _____ Hamate   _____ Trapezium
   _____ Lunate   _____ Trapezoid
   _____ Pisiform   _____ Triquetrum

4. Which carpal is the most commonly dislocated carpal?
   _____ Capitate   _____ Scaphoid
   _____ Hamate   _____ Trapezium
   _____ Lunate   _____ Trapezoid
   _____ Pisiform   _____ Triquetrum

5. The drawing below (Fig. 4-1) shows the outlines of the carpals (the bones of the wrist), the metacarpals (the bones of the palm of the hand), and the phalanges (the bones of the thumb and fingers) in an AP radiograph of the hand. Print the letter of each bone and bone part listed below over its outline in the drawing. Label each joint listed.

A. Capitate
B. Hamate
C. Lunate
D. Pisiform
E. Scaphoid
F. Trapezium
G. Trapezoid
H. Triquetrum
I. Base of 1st metacarpal
J. Head of 4th metacarpal
K. Shaft of distal phalanx of the thumb

L. Head of proximal phalanx of the index finger
M. Shaft of middle phalanx of the ring finger
N. Base of distal phalanx of the little finger
O. Styloid process of the radius
P. Styloid process of the ulna
Q. Carpometacarpal joint of the thumb
R. Metacarpophalangeal joint of the index finger
S. Distal interphalangeal joint of the middle finger
T. Proximal interphalangeal joint of the little finger

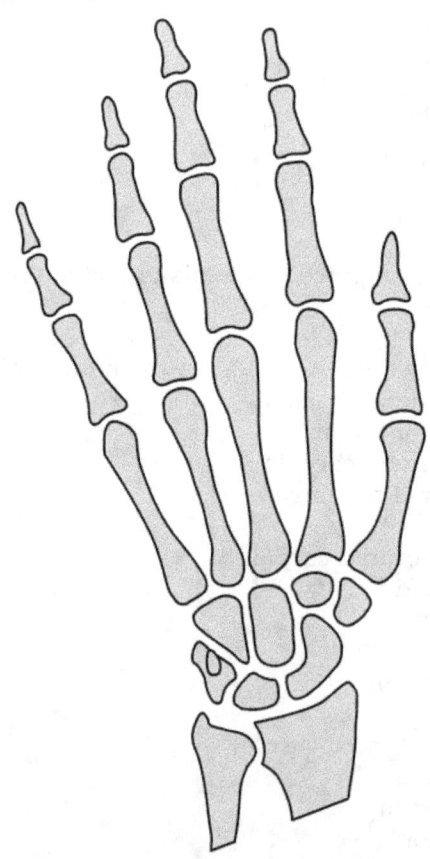

Fig. 4-1

6. The wrist joint is a synovial joint in which the distal end of the radius and the articular disc articulate with 3 carpals (the articular disc is a triangular-shaped disc of cartilage that lies immediately distal to the distal end of the ulna). Identify the 3 carpals.

_____ Capitate   _____ Scaphoid
_____ Hamate   _____ Trapezium
_____ Lunate   _____ Trapezoid
_____ Pisiform   _____ Triquetrum

7. Two skin creases called the proximal and distal wrist creases extend transversely across the anterior surface of the wrist. Which skin crease overlies the wrist joint? _____

8. The midcarpal joint is a synovial joint in which the three carpals identified in question 6 articulate with 4 of the other carpals. Identify the 4 carpals.

_____ Capitate   _____ Scaphoid
_____ Hamate   _____ Trapezium
_____ Lunate   _____ Trapezoid
_____ Pisiform   _____ Triquetrum

9. Which of the following hand movements involve movements within the wrist joint?

_____ Abduction of the hand
_____ Adduction of the hand
_____ Extension of the hand
_____ Flexion of the hand

10. Which of the following hand movements involve movements within the midcarpal joint?

_____ Abduction of the hand
_____ Adduction of the hand
_____ Extension of the hand
_____ Flexion of the hand

11. Identify the 4 movements which can occur at the metacarpophalangeal joint (MCP) of each finger.

12. Identify the 2 movements which can occur at the IP joints of each finger.

13. Which two bones articulate with each other at the thumb's carpometacarpal joint?

14. Which of the following 4 thumb movements involve movement at the thumb's carpometacarpal joint?

_____ Abduction of the thumb
_____ Adduction of the thumb
_____ Extension of the thumb
_____ Flexion of the thumb

15. Which of the following 4 thumb movements involve movement at the thumb's metacarpophalangeal joint?

_____ Abduction of the thumb
_____ Adduction of the thumb
_____ Extension of the thumb
_____ Flexion of the thumb

16. Identify the 2 movements which can occur at the IP joint of the thumb.

The 8 anterior forearm muscles are listed here from A to H. Refer to this list in answering questions 17-23.

A. Flexor carpi radialis
B. Flexor carpi ulnaris
C. Flexor digitorum profundus
D. Flexor digitorum superficialis
E. Flexor pollicis longus

F. Palmaris longus
G. Pronator quadratus
H. Pronator teres

17. _____ and _____ are the sole pronators of the forearm.

18. _____ can both flex and abduct the hand at the wrist and midcarpal joints.

19. _____ can only flex the hand at the wrist and midcarpal joints.

20. _____ can both flex and adduct the hand at the wrist and midcarpal joints.

21. _____ can flex the thumb at its carpometacarpal, metacarpophalangeal, and interphalangeal joints.

22. _____ can flex each of the fingers at their MCP, proximal IP, and distal IP joints.

23. _____ can flex each finger at its MCP and proximal IP joints but not its distal IP joint.

The 3 nerves which innervate the anterior forearm muscles are listed here from A to C. Refer to this list in matching each anterior forearm muscle with its innervation.

A. Anterior interosseous nerve
B. Median nerve
C. Ulnar nerve

24. _____ Flexor carpi radialis

25. _____ Flexor carpi ulnaris

26. _____ Lateral half of flexor digitorum profundus

27. _____ Medial half of flexor digitorum profundus

28. _____ Flexor digitorum superficialis

29. _____ Flexor pollicis longus

30. _____ Palmaris longus

31. _____ Pronator quadratus

32. _____ Pronator teres

33. The carpal tunnel is literally a tunnel-like space in the anterior part of the wrist. Name the structure which forms the roof of the carpal tunnel and the structures which form the floor and side walls of the tunnel.

34. One nerve and the insertion tendons of 3 anterior forearm muscles extend from the forearm into the hand by passing through the carpal tunnel. Identify the nerve and the 3 forearm muscles.

35. Explain why it can be said that C8 is the spinal nerve which controls flexion of the thumb and fingers at their interphalangeal joints.

36. What is Volkmann's ischemic contracture? What is the anatomical basis of its occurrence?

The 11 posterior forearm muscles are listed here from A to K. Refer to this list in answering questions 37-46.

A. Abductor pollicis longus	G. Extensor digitorum
B. Brachioradialis	H. Extensor indicis
C. Extensor carpi radialis brevis	I. Extensor pollicis brevis
D. Extensor carpi radialis longus	J. Extensor pollicis longus
E. Extensor carpi ulnaris	K. Supinator
F. Extensor digiti minimi	

37. _____ is the only forearm muscle which can supinate the forearm.

38. _____ can flex the forearm at the elbow joint when the forearm is in a midprone position.

39. _____ and _____ can both extend and abduct the hand at the wrist and midcarpal joints.

40. _____ can both extend and adduct the hand at the wrist and midcarpal joints.

41. _____ can extend all 4 fingers at their MCP and IP joints.

42. _____ can extend the index finger at its MCP and IP joints.

43. _____ can extend the little finger at its MCP and IP joints.

44. _____ can both extend and abduct the thumb at its carpometacarpal joint.

45. _____ can extend the thumb at its carpometacarpal and MCP joints.

46. _____ can extend the thumb at its carpometacarpal, MCP, and IP joints

The 3 nerves which innervate the posterior forearm muscles are listed here from A to C. Refer to this list in matching each posterior forearm muscle with its innervation.

A. Radial nerve   B. Deep radial nerve   C. Posterior interosseous nerve

47. _____ Abductor pollicis longus

48. _____ Brachioradialis

49. _____ Extensor carpi radialis brevis

50. _____ Extensor carpi radialis longus

51. _____ Extensor carpi ulnaris

52. _____ Extensor digiti minimi

53. _____ Extensor digitorum

54. _____ Extensor indicis

55. _____ Extensor pollicis brevis

56. _____ Extensor pollicis longus

57. _____ Supinator

58. What is wrist drop?

59. What is lateral epicondylitis?

60. The anatomical snuffbox is a triangular-shaped, hollowed area on the lateral side of the wrist which is best seen if both the hand is adducted at the wrist and midcarpal joints and the thumb is extended at its carpometacarpal, MCP, and IP joints. Identify the two insertion tendons that form the lateral border of the snuffbox and the single insertion tendon that forms the medial border of the snuffbox.

61. What is the anatomical basis for palpating the floor of the anatomical snuffbox when examining a patient for a possible scaphoid fracture?

62. What is a relatively common complication of scaphoid fractures?

63. What is a Colles' fracture?

64. When you answered question 23 in the chapter on the axilla, you identified that there are 4 shoulder which receive all of their innervation from only C5 and C6 nerve fibers: deltoid and three of the rotator cuff muscles, namely, supraspinatus, infraspinatus, and teres minor. When you answered question 15 in the chapter on the arm, you identified biceps brachii and brachialis as the two arm muscles which receive most of their innervation from C5 and C6 nerve fibers. In the forearm, there are two posterior forearm muscles which receive all their innervation from only C5 and C6 nerve fibers: brachioradialis and supinator.

When a person suffers an upper brachial plexus injury, the nerve fibers which are generally affected the most are C5 and C6 nerve fibers. Therefore, upper brachial plexus injuries most commonly result in weakness or paralysis of those upper limb muscles which receive most or all of their innervation from C5 and C6 nerve fibers, namely, the 8 muscles identified above. The motor deficits which result if all the C5 and C6 nerve fibers are destroyed are collectively called Erb's palsy. Given your knowledge of the actions of the 8 muscles identified above, describe the motor deficits of a person suffering from Erb's palsy.

Skin of the Forearm

65. Most of the skin on the lateral side of the forearm is part of a dermatome. This is the dermatome of _____.

_____ C1
_____ C2
_____ C3
_____ C4
_____ C5
_____ C6
_____ C7
_____ C8
_____ T1
_____ T2
_____ T3
_____ T4
_____ T5

66. Most of the skin on the medial side of the forearm is part of a dermatome. This is the dermatome of _____.

_____ C1
_____ C2
_____ C3
_____ C4
_____ C5
_____ C6
_____ C7
_____ C8
_____ T1
_____ T2
_____ T3
_____ T4
_____ T5

Peripheral Pulses at the Wrist

67. Where is the radial artery pulse palpable near the wrist?

68. Where is the ulnar artery pulse palpable near the wrist?

END OF QUESTIONS IN PART A OF THE CHAPTER ON THE FOREARM AND WRIST

FOREARM AND WRIST – Part B: Questions and Answers

Dissection of the forearm and wrist in gross lab focuses on identification of the anterior and posterior forearm muscles, the flexor and extensor retinacula, the median, anterior interosseous, deep radial, posterior interosseous, and ulnar nerves, and the radial and ulnar arteries. Five topics dominate the forearm anatomy most frequently applied in clinical practice: (1) the use of the radial and ulnar arteries at the wrists as the best sites to measure a patient's pulse, (2) knowledge of the courses by which the median and ulnar nerves extend from the forearm into the hand, (3) the use of the anatomical snuffbox for the examination of a fractured scaphoid, (4) the nature of a Colles' fracture, and (5) the innervation and major actions of the muscles of the forearm.

Bones & Joints Associated with the Actions of Forearm Muscles

The muscles of the forearm can move the forearm at the proximal and distal radioulnar joints, the hand at the wrist and midcarpal joints, and the thumb and fingers at their metacarpophalangeal (MCP) and interphalangeal (IP) joints.

In the following pages, as you answer questions related to the bones and joints of the hand, it is important to recognize that anatomically and medically the thumb is not a finger. In the hand, there are five digits, and they consist of the four fingers plus the thumb. Observe that each digit has two names. When you identify a digit of the hand in clinical notes, it is advisable to include both names of the digit:

Thumb-1st digit
Index finger-2nd digit
Middle finger-3rd digit
Ring finger-4th digit
Little finger-5th digit

1. The styloid process of the radius can be palpated at the distal end of the radius:
    <u>True</u> or False

2. When a person makes a fist, which bony surfaces underlie the knuckles?
    _____ Bases of the 2nd, 3rd, 4th and 5th metacarpals
    _____ Bases of the proximal phalanges of the fingers
    __x__ Heads of the 2nd, 3rd, 4th, and 5th metacarpals
    _____ Heads of the proximal phalanges of the fingers

3. Which carpal is the most commonly fractured carpal?
    _____ Capitate   __x__ Scaphoid
    _____ Hamate   _____ Trapezium
    _____ Lunate   _____ Trapezoid
    _____ Pisiform   _____ Triquetrum

4. Which carpal is the most commonly dislocated carpal?
    _____ Capitate   _____ Scaphoid
    _____ Hamate   _____ Trapezium
    __x__ Lunate   _____ Trapezoid
    _____ Pisiform   _____ Triquetrum

5. The drawing below (Fig. 4-1) shows the outlines of the carpals (the bones of the wrist), the metacarpals (the bones of the palm of the hand, and the phalanges (the bones of the thumb and fingers) in an AP radiograph of the hand. Print the letter of each bone and bone part listed below over its outline in the drawing. Label each joint listed.

A. Capitate
B. Hamate
C. Lunate
D. Pisiform
E. Scaphoid
F. Trapezium
G. Trapezoid
H. Triquetrum
I. Base of 1st metacarpal
J. Head of 4th metacarpal
K. Shaft of distal phalanx
   of the thumb

L. Head of proximal phalanx of the index finger
M. Shaft of middle phalanx of the ring finger
N. Base of distal phalanx of the little finger
O. Styloid process of the radius
P. Styloid process of the ulna
Q. Carpometacarpal joint of the thumb
R. Metacarpophalangeal joint of the index finger
S. Distal interphalangeal joint of
   the middle finger
T. Proximal interphalangeal joint of
   the little finger

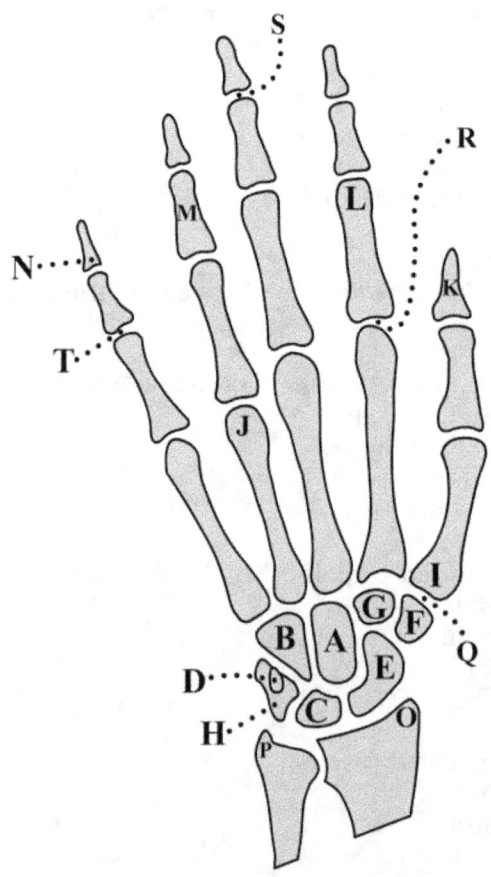

Fig. 4-1

6. The wrist joint is a synovial joint in which the distal end of the radius and the articular disc articulate with 3 carpals (the articular disc is a triangular-shaped disc of cartilage that lies immediately distal to the distal end of the ulna). Identify the 3 carpals.

____ Capitate   __x_ Scaphoid
____ Hamate   ____ Trapezium
__x_ Lunate   ____ Trapezoid
____ Pisiform   __x_ Triquetrum

7. Two skin creases called the proximal and distal wrist creases extend transversely across the anterior surface of the wrist. Which skin crease overlies the wrist joint?
__the proximal wrist crease__

8. The midcarpal joint is a synovial joint in which the three carpals identified in question 6 articulate with 4 of the other carpals. Identify the 4 carpals.

__x_ Capitate   ____ Scaphoid
__x_ Hamate   __x_ Trapezium
____ Lunate   __x_ Trapezoid
____ Pisiform   ____ Triquetrum

9. Which of the following hand movements involve movements within the wrist joint?
__x_ Abduction of the hand
__x_ Adduction of the hand
__x_ Extension of the hand
__x_ Flexion of the hand

10. Which of the following hand movements involve movements within the midcarpal joint?
__x_ Abduction of the hand
__x_ Adduction of the hand
__x_ Extension of the hand
__x_ Flexion of the hand

11. Identify the 4 movements which can occur at the metacarpophalangeal joint (MCP) of each finger.
__Abduction of the finger__
__Adduction of the finger__
__Extension of the finger's proximal phalanx at the MCP joint__
__Flexion of the finger's proximal phalanx at the MCP joint__

12. Identify the 2 movements which can occur at the IP joints of each finger.
__Extension of the more distal phalanx at each IP joint__
__Flexion of the more distal phalanx at each IP joint__

13. Which two bones articulate with each other at the thumb's carpometacarpal joint?
__Trapezium__
__1st Metacarpal__

14. Which of the following 4 thumb movements involve movement at the thumb's carpometacarpal joint?
__x_ Abduction of the thumb
__x_ Adduction of the thumb
__x_ Extension of the thumb
__x_ Flexion of the thumb

15. Which of the following 4 thumb movements involve movement at the thumb's metacarpophalangeal joint?
__x__ Abduction of the thumb
__x__ Adduction of the thumb
__x__ Extension of the thumb
__x__ Flexion of the thumb

16. Identify the 2 movements which can occur at the IP joint of the thumb.
__Extension of the distal phalanx__
__Flexion of the distal phalanx__

Muscles of the Forearm

The drawing below (Fig. 4-2) shows a transverse sectional view of the right forearm's midregion. Deep to the skin lie two layers of fascia: a layer of superficial fascia and a thin layer of deep fascia. Notice that all the forearm muscles lie deep to the deep fascia layer, which, in the forearm, is called the antebrachial fascia. The antebrachial fascia gives rise to a sheet-like layer called an intermuscular septum that extends deeply to attach to the shaft of the radius. The radius and ulna are attached to each other along their shafts by the interosseous membrane of the forearm. The antebrachial fascia and its intermuscular septum in association with the radius, ulna, and interosseous membrane divide the forearm into two muscular compartments: an anterior compartment and a posterior compartment. The muscles of the forearm are thus divisible into two large groups: the anterior forearm muscles and the posterior forearm muscles.

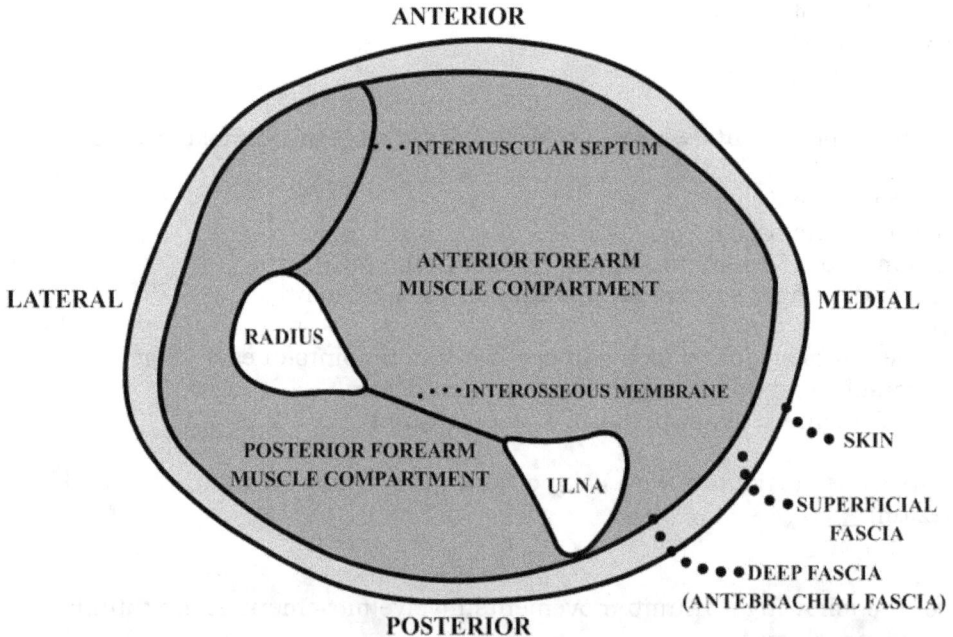

Fig. 4-2: Transverse cross section of the right forearm's midregion

The 8 anterior forearm muscles are listed here from A to H. Refer to this list in answering questions 17-23.

A. Flexor carpi radialis
B. Flexor carpi ulnaris
C. Flexor digitorum profundus
D. Flexor digitorum superficialis
E. Flexor pollicis longus

F. Palmaris longus
G. Pronator quadratus
H. Pronator teres

17. __G__ and __H__ are the sole pronators of the forearm.

18. __A__ can both flex and abduct the hand at the wrist and midcarpal joints.

19. __F__ can only flex the hand at the wrist and midcarpal joints.

20. __B__ can both flex and adduct the hand at the wrist and midcarpal joints.

21. __E__ can flex the thumb at its carpometacarpal, metacarpophalangeal, and interphalangeal joints.

22. __C__ can flex each of the fingers at their MCP, proximal IP, and distal IP joints.

23. __D__ can flex each finger at its MCP and proximal IP joints but not its distal IP joint.

The 3 nerves which innervate the anterior forearm muscles are listed here from A to C. Refer to this list in matching each anterior forearm muscle with its innervation.

A. Anterior interosseous nerve
B. Median nerve
C. Ulnar nerve

24. __B__ Flexor carpi radialis

25. __C__ Flexor carpi ulnaris

26. __A__ Lateral half of flexor digitorum profundus

27. __C__ Medial half of flexor digitorum profundus

28. __B__ Flexor digitorum superficialis

29. __A__ Flexor pollicis longus

30. __B__ Palmaris longus

31. __A__ Pronator quadratus

32. __B__ Pronator teres

33. The carpal tunnel is literally a tunnel-like space in the anterior part of the wrist. Name the structure which forms the roof of the carpal tunnel and the structures which form the floor and side walls of the tunnel.

   Roof __flexor retinaculum__

   Floor and side walls __the carpals__

34. One nerve and the insertion tendons of 3 anterior forearm muscles extend from the forearm into the hand by passing through the carpal tunnel. Identify the nerve and the 3 forearm muscles.

   __median nerve__insertion tendon of flexor pollicis longus__insertion tendons of flexor digitorum superficialis__insertion tendons of flexor digitorum profundus__

35. Explain why it can be said that C8 is the spinal nerve which controls flexion of the thumb and fingers at their interphalangeal joints.

C8 is the spinal nerve which provides most of the nerve fibers to the only muscle which flexes the thumb at its IP joint (flexor pollicis longus) and the only muscles which flex the fingers at their IP joints (flexor digitorum profundus and flexor digitorum superficialis). Therefore, C8 is the spinal nerve which provides most of the nerve fibers that control flexion of the thumb and fingers at their IP joints.

36. What is Volkmann's ischemic contracture? What is the anatomical basis of its occurrence?

As illustrated in Fig. 4-2, the anterior and posterior groups of forearm muscles are each surrounded by a combination of the forearm's deep fascia, an intermuscular septum, the shafts of the radius and ulna, and the interosseous membrane of the forearm. Each group of forearm muscles is thus enveloped within a closed space, or compartment, bordered by bony surfaces and relatively inelastic, connective tissue layers. This enclosure of the anterior and posterior forearm muscle groups within closed spaces of fixed volume puts each muscle group at risk of ischemia (that is, loss of blood supply) if injury or disease should significantly increase the pressure within the compartment. The medical term compartment syndrome is used to refer to any condition that leads to compression of blood vessels, nerves, and muscles in muscle compartments of the limbs. The anterior and posterior muscle compartments of the forearm and the anterior, lateral, deep posterior, and superficial posterior muscle compartments of the leg are the limb compartments most sensitive to compartment syndrome.

Compression of the anterior forearm muscle compartment is a common complication of distal humeral and radial or ulnar shaft fractures. It is especially noteworthy that supracondylar fractures of the humerus (fractures immediately proximal to the distal end of the humerus) are common in children, frequently occurring as a result of a fall on an outstretched hand. A child with a supracondylar humeral fracture is always at risk that untreated swelling of the anterior forearm muscle compartment will cause ischemic necrosis (that is, death by loss of blood supply) of anterior forearm muscles. If such ischemic necrosis occurs, the necrotic muscle tissue is replaced by fibrous scar tissue that permanently shortens the affected anterior forearm muscles and draws the thumb and fingers into marked flexion. This deformity is called Volkmann's ischemic contracture.

The 11 posterior forearm muscles are listed here from A to K. Refer to this list in answering questions 37-46.

A. Abductor pollicis longus
B. Brachioradialis
C. Extensor carpi radialis brevis
D. Extensor carpi radialis longus
E. Extensor carpi ulnaris
F. Extensor digiti minimi

G. Extensor digitorum
H. Extensor indicis
I. Extensor pollicis brevis
J. Extensor pollicis longus
K. Supinator

37. __K__ is the only forearm muscle which can supinate the forearm.

38. __B__ can flex the forearm at the elbow joint when the forearm is in a midprone position.

39. __C__ and __D__ can both extend and abduct the hand at the wrist and midcarpal joints.

40. __E__ can both extend and adduct the hand at the wrist and midcarpal joints.

41. __G__ can extend all 4 fingers at their MCP and IP joints.

42. __H__ can extend the index finger at its MCP and IP joints.

43. __F__ can extend the little finger at its MCP and IP joints.

44. __A__ can both extend and abduct the thumb at its carpometacarpal joint.

45. __I__ can extend the thumb at its carpometacarpal and MCP joints.

46. __J__ can extend the thumb at its carpometacarpal, MCP, and IP joints

The 3 nerves which innervate the posterior forearm muscles are listed here from A to C. Refer to this list in matching each posterior forearm muscle with its innervation.

A. Radial nerve   B. Deep radial nerve   C. Posterior interosseous nerve

47. __C__ Abductor pollicis longus

48. __A__ Brachioradialis

49. __B__ Extensor carpi radialis brevis

50. __A__ Extensor carpi radialis longus

51. __C__ Extensor carpi ulnaris

52. __C__ Extensor digiti minimi

53. __C__ Extensor digitorum

54. __C__ Extensor indicis

55. __C__ Extensor pollicis brevis

56. __C__ Extensor pollicis longus

57. __B__ Supinator

58. What is wrist drop?

Wrist drop is the condition in which a person is unable to extend the hand at the wrist and midcarpal joints. Wrist drop is occasionally the result of an inebriated person falling asleep in a chair with his/her arm draped over the back of the chair. The resulting pressure on the radial nerve in the axilla may lead to a temporary case of wrist drop. Wrist drop acquired under such a condition is sometimes referred to as 'Saturday night palsy.'

59. What is lateral epicondylitis?

Extensor carpi radialis brevis, extensor carpi ulnaris, extensor digitorum, and extensor digiti minimi all originate from the lateral epicondyle of the humerus via a common tendon of origin called the common extensor tendon. Repetitive, strenuous exercise of these muscles (such as may occur during backhand strokes in a tennis match) may lead to degeneration of the common extensor tendon; such a condition is called lateral epicondylitis. The condition is frequently called tennis elbow because of its relatively high incidence among tennis players. The patient typically presents with pain in the lateral aspect of the elbow that may radiate down the back of the forearm into the dorsum of the hand. Physical examination reveals that the point of maximum tenderness in the lateral aspect of the elbow region is that point which overlies the anterior aspect of the lateral epicondyle of the humerus. The pain may be exacerbated by requesting the patient to lie supine with the forearm fully extended (at the elbow joint) and fully pronated (at the radioulnar joints) and then to attempt extension of the hand at the wrist against resistance provided by the examiner. This isometric exercise exacerbates the elbow pain by exerting tension on the anterior part of the common extensor tendon.

60. The anatomical snuffbox is a triangular-shaped, hollowed area on the lateral side of the wrist which is best seen if both the hand is adducted at the wrist and midcarpal joints and the thumb is extended at its carpometacarpal, MCP, and IP joints. Identify the two insertion tendons that form the lateral border of the snuffbox and the single insertion tendon that forms the medial border of the snuffbox.
   Lateral border __insertion tendons of abductor pollicis longus and
                         extensor pollicis brevis__
   Medial border __insertion tendon of extensor pollicis longus__

61. What is the anatomical basis for palpating the floor of the anatomical snuffbox when examining a patient for a possible scaphoid fracture?

Four bones or bony parts form the floor of the anatomical snuffbox. Proceeding from the proximal end of the snuffbox to its distal end, the bony floor consists of the styloid process of the radius, the scaphoid, the trapezium, and the base of the 1st metacarpal. Consequently, pressure upon the floor of the snuffbox will intensify a patient's wrist pain if the scaphoid is fractured.

62. What is a relatively common complication of scaphoid fractures?

Scaphoid fractures are at risk of non-union of the fragments. This is because most or all of the scaphoid's nutrient foramina are located on the surface of the distal half of the bone. If the fracture tears apart inner arteries and these inner arteries fail to reunite, the proximal fragment of the bone may suffer ischemic necrosis. The more proximal the location of the fracture site, the greater the risk of avacular necrosis of the proximal fragment.

Many scaphoid fractures are hairline, undisplaced fractures which are not visible in radiographic views of the wrist. In cases in which a scaphoid fracture is strongly suspected but not radiographically confirmed, the wrist is immobilized in a plaster cast (to provide the appropriate conditions for union of the proximal and distal fragments of the scaphoid), and radiographic evaluation is repeated 7 to 10 days later. The subsequent bone resorption that occurs at the fracture site renders the site more radiolucent and thus more apparent in the latter set of radiographs.

63. What is a Colles' fracture?

A Colle's fracture is a type of distal radius fracture that frequently occurs in adults 50 years of age and older when they fall on an outstretched hand. The fracture typically extends transversely through the distal radius and results in the distal fragment being both posteriorly displaced and posteriorly angulated. Osteoporosis contributes to the relatively high prevalence of Colles' fractures among adults 50 years of age and older.

64. When you answered question 23 in the chapter on the axilla, you identified that there are 4 shoulder which receive all of their innervation from only C5 and C6 nerve fibers: deltoid and three of the rotator cuff muscles, namely, supraspinatus, infraspinatus, and teres minor. When you answered question 15 in the chapter on the arm, you identified biceps brachii and brachialis as the two arm muscles which receive most of their innervation from C5 and C6 nerve fibers. In the forearm, there are two posterior forearm muscles which receive all their innervation from only C5 and C6 nerve fibers: brachioradialis and supinator.

When a person suffers an upper brachial plexus injury, the nerve fibers which are generally affected the most are C5 and C6 nerve fibers. Therefore, upper brachial plexus injuries most commonly result in weakness or paralysis of those upper limb muscles which receive most or all of their innervation from C5 and C6 nerve fibers, namely, the 8 muscles identified above. The motor deficits which result if all the C5 and C6 nerve fibers are destroyed are collectively called Erb's palsy. Given your knowledge of the actions of the 8 muscles identified above, describe the motor deficits of a person suffering from Erb's palsy.

First, the person is unable to raise their arm at the shoulder joint because the only two muscles which can abduct the arm at the shoulder joint, supraspinatus and deltoid, are both paralyzed. Second, the person cannot externally rotate the arm at the shoulder joint because all the muscles which can externally rotate the arm at the shoulder joint (infraspinatus, teres minor, and deltoid) are all paralyzed. Third, the person can barely flex the forearm at the elbow joint because the two prime movers of forearm flexion, biceps brachii and brachialis, are greatly weakened and an accessory forearm flexor, brachioradialis, is completely paralyzed. Fourth, and finally, the person can barely supinate the forearm because the only two muscles which can supinate the forearm, biceps brachii and supinator, are either greatly weakened or completely paralyzed.

Skin of the Forearm

65. Most of the skin on the lateral side of the forearm is part of a dermatome. This is the dermatome of _____.

```
_____ C1
_____ C2
_____ C3
_____ C4
_____ C5
__x__ C6
_____ C7
_____ C8
_____ T1
_____ T2
_____ T3
_____ T4
_____ T5
```

66. Most of the skin on the medial side of the forearm is part of a dermatome. This is the dermatome of _____.

```
_____ C1
_____ C2
_____ C3
_____ C4
_____ C5
_____ C6
_____ C7
__x__ C8
_____ T1
_____ T2
_____ T3
_____ T4
_____ T5
```

Peripheral Pulses at the Wrist

67. Where is the radial artery pulse palpable near the wrist?

Pulsations of the radial artery can be palpated on the anterolateral side of the forearm 2 to 3 cm proximal to the distal crease of the wrist.

68. Where is the ulnar artery pulse palpable near the wrist?

Pulsations of the ulnar artery can be palpated on the anteromedial side of the forearm about 2 cm proximal to the distal crease of the wrist.

HAND – Part A: Questions

Dissection of the hand in gross lab focuses on identification of the 18 intrinsic hand muscles, the relationship between the flexor digitorum profundus and superficialis tendons in the fingers, branches of the median nerve and ulnar nerve in the hand, and the superficial and deep palmar arches. Four topics dominate the hand anatomy most frequently applied in clinical practice: (1) knowledge of the cutaneous sensory innervation provided by the median, ulnar, and superficial radial nerves, (2) knowledge of the dermatomes at the tips of the thumb, middle finger, and little finger, (3) testing of motor nerve supply to the intrinsic hand muscles, and (4) knowledge of the disfigurement resulting from (a) paralysis of the muscles of the thenar eminence, (b) paralysis of the intrinsic hand muscles innervated by ulnar nerve fibers, and (c) a severe lower brachial plexus injury.

Intrinsic Hand Muscles

The muscles which lie exclusively in the hand are called the intrinsic hand muscles. Individual and groups of intrinsic hand muscles are listed here from A to J. Refer to this list in answering questions 1-10.

A. Abductor digiti minimi   E. Flexor digiti minimi   I. Opponens pollicis
B. Abductor pollicis brevis   F. Flexor pollicis brevis   J. Palmar interossei
C. Adductor pollicis   G. Lumbricals
D. Dorsal interossei   H. Opponens digiti minimi

1. _____ is the only muscle which can oppose the thumb.

2. _____ can abduct the thumb at its metacarpophalangeal joint.

3. _____ can adduct the thumb at its metacarpophalangeal joint.

4. _____ can flex the thumb at its metacarpophalangeal joint.

5. The _____ can flex the fingers at their MCP joints and extend the fingers at their IP joints.

6. The _____ can abduct the index, middle, and ring fingers at their MCP joints.

7. The _____ can adduct the index, ring, and little fingers at their MCP joints.

8. _____ can abduct the little finger at its MCP joint.

9. _____ can flex the little finger at its MCP joint.

10. _____ can laterally rotate the 5th metacarpal.

11. Each intrinsic hand muscle is innervated either by the median nerve or the deep branch of the ulnar nerve. What is the shortest description of the innervation of the intrinsic hand muscles?

12. What is ape hand?

13. How do the fingers become disfigured following paralysis of their lumbricals?

14. What is claw hand?

15. Which two spinal nerves provide all the motor and sensory nerve fibers to the intrinsic hand muscles?

_____ C5   _____ C6   _____ C7   _____ C8   _____ T1

16. As noted by the answer to the previous question, every intrinsic hand muscle, independently of whether it is innervated by the median nerve or the deep branch of the ulnar nerve, receives all of its nerve fibers from only two spinal nerves: C8 and T1. When a person suffers a lower brachial plexus injury, the nerve fibers which are generally affected the most are C8 and T1 nerve fibers. Therefore, lower brachial plexus injuries most commonly result in weakness or paralysis of those upper limb muscles which receive all of their innervation from C8 and T1 nerve fibers, namely, all the intrinsic hand muscles. The hand disfigurement which results if all the C8 and T1 nerve fibers are destroyed is called Klumpke's palsy. Given your answers to questions 12, 13, and 14, describe the hand disfigurement of a person suffering from Klumpke's palsy.

17. Explain why it can be said that T1 is the spinal nerve which controls abduction and adduction of the fingers at their metacarpophalangeal joints.

18. Fig. 5-1 shows outline drawings of the palmar and dorsal surfaces of the hand and its digits. Use a pencil to demarcate the cutaneous areas innervated by median, ulnar, and superficial radial nerve fibers.

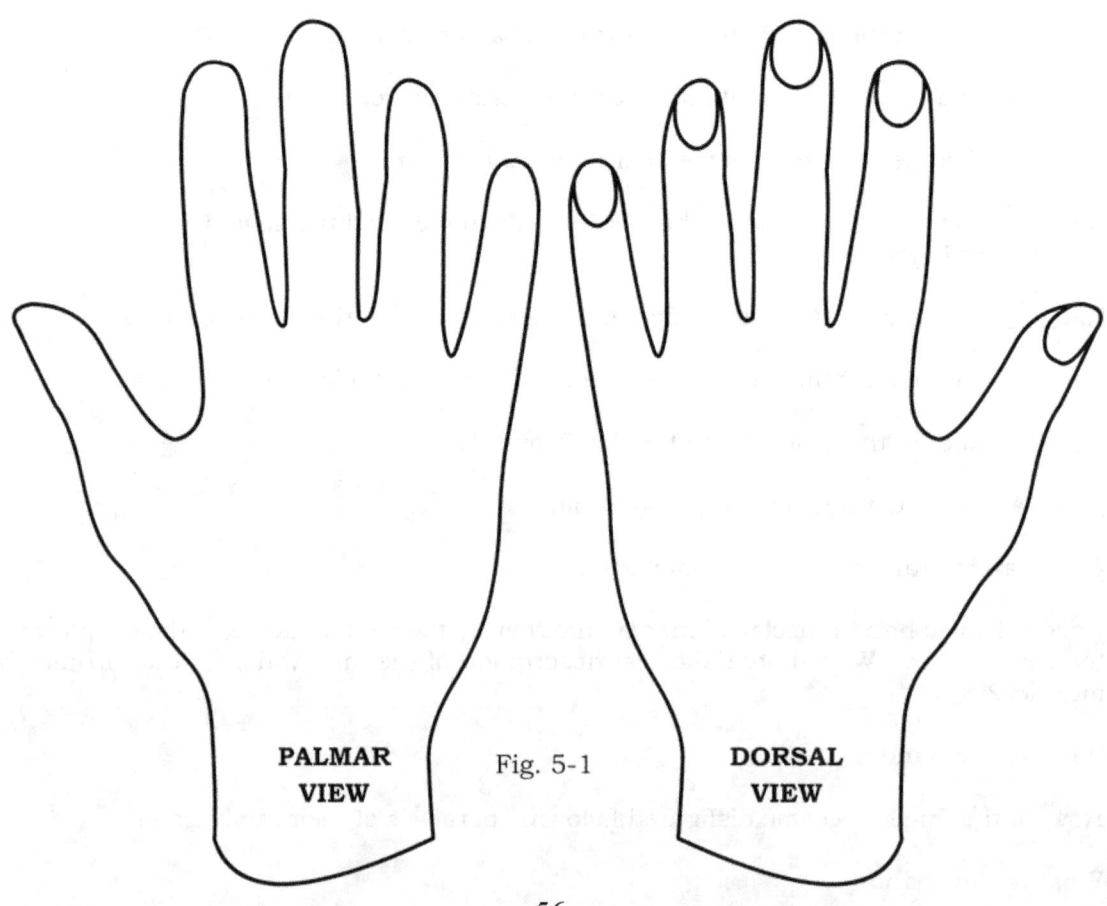

PALMAR VIEW   Fig. 5-1   **DORSAL VIEW**

19. What are the symptoms of carpal tunnel syndrome?

20. Three spinal nerves (C6, C7, and C8) provide almost all the cutaneous nerve fibers for the palmar surface of the hand and its thumb and fingers. The anterior surface at the tip of one of the hand's digits is part of the C6 dermatome, a second is part of the C7 dermatome, and a third is part of the C8 dermatome. In the drawing of Fig. 5-2, mark the digits whose tips are parts of the C6, C7, and C8 dermatomes.

21. Opponens pollicis is the sole prime mover of opposition of the thumb, which is the movement by which the tip of the thumb's palmar surface, starting from the anatomical position, is brought into contact with the tips of the palmar surfaces of the fingers. Opponens pollicis also makes it possible for the tip of a person's thumb to be brought into contact with a specific, small region on the surface of the palm. In the drawing of Fig. 5-3, mark the region on the surface of the palm which can be touched by the tip of the thumb if and only if opponens pollicis is fully functional.

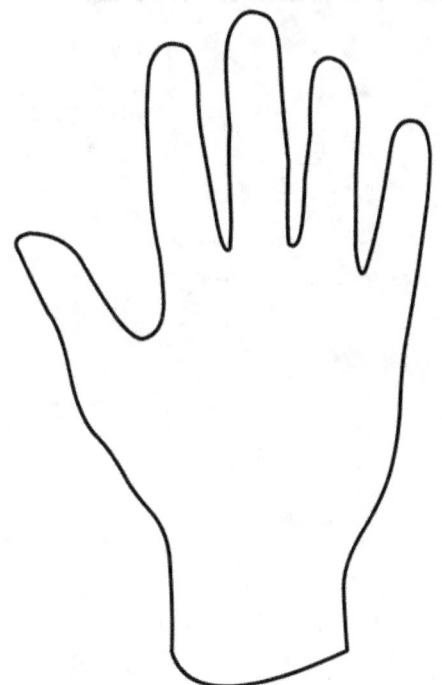

Fig. 5-2
Outline of the palmar surface
of the hand

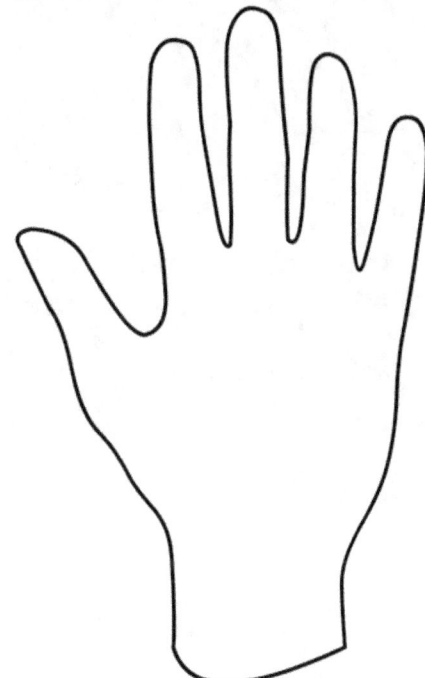

Fig. 5-3
Outline of the palmar surface
of the hand

22. During an examination of an injured person who has been delivered by ambulance to the ER, the physician inserts a piece of paper between the patient's index and middle fingers and asks the patient to resist the attempt by the physician to remove the piece of paper. The physician employs this test to assess the integrity of a major nerve of the upper limb. Identify the nerve and explain how this test assesses its integrity.

23. There are two arched arteries in the palm of the hand called the superficial and deep palmar arches. The superficial palmar arch lies anterior to the flexor digitorum tendons in the palm of the hand, and the deep palmar arch lies posterior to these tendons. In most persons, both the radial and ulnar arteries supply both arterial arches. Describe the Allen test and explain its purpose.

24. What is the anatomical and physiological basis for the fact that, in a normal Allen test, a blood capillary bed in the palm of the hand that has been depleted of its blood supply should become filled with blood within a 2 to 4 second interval after blood supply has been fully restored?

END OF QUESTIONS IN PART A OF THE CHAPTER ON THE HAND

HAND – Part B: Questions and Answers

Dissection of the hand in gross lab focuses on identification of the 18 intrinsic hand muscles, the relationship between the flexor digitorum profundus and superficialis tendons in the fingers, branches of the median nerve and ulnar nerve in the hand, and the superficial and deep palmar arches. Four topics dominate the hand anatomy most frequently applied in clinical practice: (1) knowledge of the cutaneous sensory innervation provided by the median, ulnar, and superficial radial nerves, (2) knowledge of the dermatomes at the tips of the thumb, middle finger, and little finger, (3) testing of motor nerve supply to the intrinsic hand muscles, and (4) knowledge of the disfigurement resulting from (a) paralysis of the muscles of the thenar eminence, (b) paralysis of the intrinsic hand muscles innervated by ulnar nerve fibers, and (c) a severe lower brachial plexus injury.

Intrinsic Hand Muscles

The muscles which lie exclusively in the hand are called the intrinsic hand muscles. Individual and groups of intrinsic hand muscles are listed here from A to J. Refer to this list in answering questions 1-10.

A. Abductor digiti minimi   E. Flexor digiti minimi   I. Opponens pollicis
B. Abductor pollicis brevis F. Flexor pollicis brevis   J. Pamar inteorssei
C. Adductor pollicis   G. Lumbricals
D. Dorsal interossei   H. Opponens digit minimi

1. __I__ is the only muscle which can oppose the thumb.

2. __B__ can abduct the thumb at its metacarpophalangeal joint.

3. __C__ can adduct the thumb at its metacarpophalangeal joint.

4. __F__ can flex the thumb at its metacarpophalangeal joint.

5. The __G__ can flex the fingers at their MCP joints and extend the fingers at their IP joints.

6. The __D__ can abduct the index, middle, and ring fingers at their MCP joints.

7. The __J__ can adduct the index, ring, and little fingers at their MCP joints.

8. __A__ can abduct the little finger at its MCP joint.

9. __E__ can flex the little finger at its MCP joint.

10. __H__ can laterally rotate the 5th metacarpal.

11. Each intrinsic hand muscle is innervated either by the median nerve or the deep branch of the ulnar nerve. What is the shortest description of the innervation of the intrinsic hand muscles?

The median nerve innervates only 5 intrinsic hand muscles: the three muscles of the thenar eminence (abductor pollicis brevis, flexor pollicis brevis, and opponens pollicis) and the 1st and 2nd lumbricals. The remaining 13 intrinsic hand muscles are all innervated by the deep branch of the ulnar nerve.

12. What is ape hand?

The thenar eminence is the mound of soft tissue in the palm at the base of the thumb. The three intrinsic hand muscles (abductor pollicis brevis, flexor pollicis brevis, and opponens pollicis) which form most of the soft tissue of the thenar eminence are called the muscles of the thenar eminence. The muscles of the thenar eminence are all innervated by a branch of the median nerve, called the recurrent branch of the median nerve, that arises from the median nerve distal to its passage through the carpal tunnel.

Ape hand is the condition which results if an injury or condition results in the loss of all or almost all the nerve fibers in the recurrent branch of the median nerve. Such a loss results in an inability to oppose the thumb because of paralysis of opponens pollicis and weakness in abduction of the thumb because of paralysis of abductor pollicis brevis. The prominence of the thenar eminence in the palm of the hand diminishes as the muscles of the thenar eminence progressively atrophy following paralysis. The thumb becomes laterally rotated excessively at its carpometacarpal joint because of the paralysis of opponens pollicis, one of whose actions during opposition of the thumb is to medially rotate the thumb at its carpometacarpal joint. The flattening of the thenar eminence due to the atrophy of the underlying muscles and the excessive lateral rotation of the thumb at its carpometacarpal joint cause the affected hand to resemble that of an ape. Consequently, paralysis of the muscles of the thenar eminence is said to produce an ape hand.

13. How do the fingers become disfigured following paralysis of their lumbricals?

Under normal resting conditions, the combined extensor activity of the 3 posterior forearm muscles which can extend 1 or all 4 four fingers (namely, extensor digitorum, extensor indicis, and extensor digiti minimi) is counterbalanced by the combined flexor activity of the lumbricals and the 2 anterior forearm muscles which can flex the fingers (namely, flexor digitorum profundus and superficialis). Paralysis of the lumbricals, however, results in the combined extensor activity of the 3 posterior forearm finger extensors being greater than the combined flexor activity of the 2 anterior forearm finger flexors. As a result of this imbalance, the fingers become hyperextended at their metacarpophalangeal joints. The hyperextension of each finger at its metacarpophalangeal joint stretches the flexor digitorum profundus and superficialis tendons in the finger to the extent that the tendons can no longer accommodate full extension of the finger at its interphalangeal joints. Consequently, the finger also becomes fully flexed at its interphalangeal joints. Paralysis of a finger's lumbrical thus results in hyperextension of the finger at its MCP joint and full flexion at its IP joints.

14. What is claw hand?

Claw hand is the condition which results if an injury causes the loss of all or almost all the nerve fibers in the ulnar nerve responsible for the innervation of intrinsic hand muscles. Such an injury results in the paralysis of 13 intrinsic hand muscles: adductor pollicis, the 3rd and 4th lumbricals, all the palmar and dorsal interossei, abductor digiti minimi, flexor digiti minimi, and opponens digiti minimi. Paralysis of the dorsal interossei and abductor digiti minimi restricts the fingers from being spread apart through abduction at their MCP joints. Paralysis of the 3rd and 4th lumbricals results in hyperextension of the ring and little fingers at their MCP joints and flexion of these fingers at their IP joints. The index and middle fingers are spared hyperextension at their MCP joints and flexion at their IP joints because the 1st and 2nd lumbricals are innervated by the median nerve. Because the disfigured ring and little fingers assume the shape of a claw, paralysis of the intrinsic hand muscles innervated by ulnar nerve fibers is said to produce a claw hand.

15. Which two spinal nerves provide all the motor and sensory nerve fibers to the intrinsic hand muscles?

_____ C5   _____ C6   _____ C7   _x_ C8   _x_ T1

16. As noted by the answer to the previous question, every intrinsic hand muscle, independently of whether it is innervated by the median nerve or the deep branch of the ulnar nerve, receives all of its nerve fibers from only two spinal nerves: C8 and T1. When a person suffers a lower brachial plexus injury, the nerve fibers which are generally affected the most are C8 and T1 nerve fibers. Therefore, lower brachial plexus injuries most commonly result in weakness or paralysis of those upper limb muscles which receive all of their innervation from C8 and T1 nerve fibers, namely, all the intrinsic hand muscles. The hand disfigurement which results if all the C8 and T1 nerve fibers are destroyed is called Klumpke's palsy. Given your answers to questions 12, 13, and 14, describe the hand disfigurement of a person suffering from Klumpke's palsy.

As noted in the answer to question 12, the thenar eminence will flatten due to the atrophy of the underlying, paralyzed thenar muscles, and the thumb will become excessively laterally rotated at its carpometacarpal joint due to the paralysis of opponens pollicis. As noted in the answer to question 13, all 4 fingers will become hyperextended at their MCP joints and fully flexed at their IP joints due to the paralysis of the lumbricals. As discussed in the answer to question 14, all 4 fingers will be adducted at their MCP joints due to the paralysis of the dorsal interossei and abductor digiti minimi. The disfigurement of Klumpke's palsy thus combines the disfigurement of ape hand and claw hand with the additional hyperextension of the index and middle fingers at their MCP joints and full flexion at their IP joints.

17. Explain why it can be said that T1 is the spinal nerve which controls abduction and adduction of the fingers at their metacarpophalangeal joints.

T1 is the spinal nerve which provides most of the nerve fibers to the only muscles which abduct the fingers (the dorsal interossei and abductor digiti minimi) and the only muscles which adduct the fingers (the palmar interossei) at their MCP joints. Therefore, T1 is the spinal nerve which provides most of the nerve fibers that control abduction and adduction of the fingers at their MCP joints.

18. Fig. 5-1 shows outline drawings of the palmar and dorsal surfaces of the hand and its digits. Use a pencil to demarcate the cutaneous areas innervated by median, ulnar, and superficial radial nerve fibers.

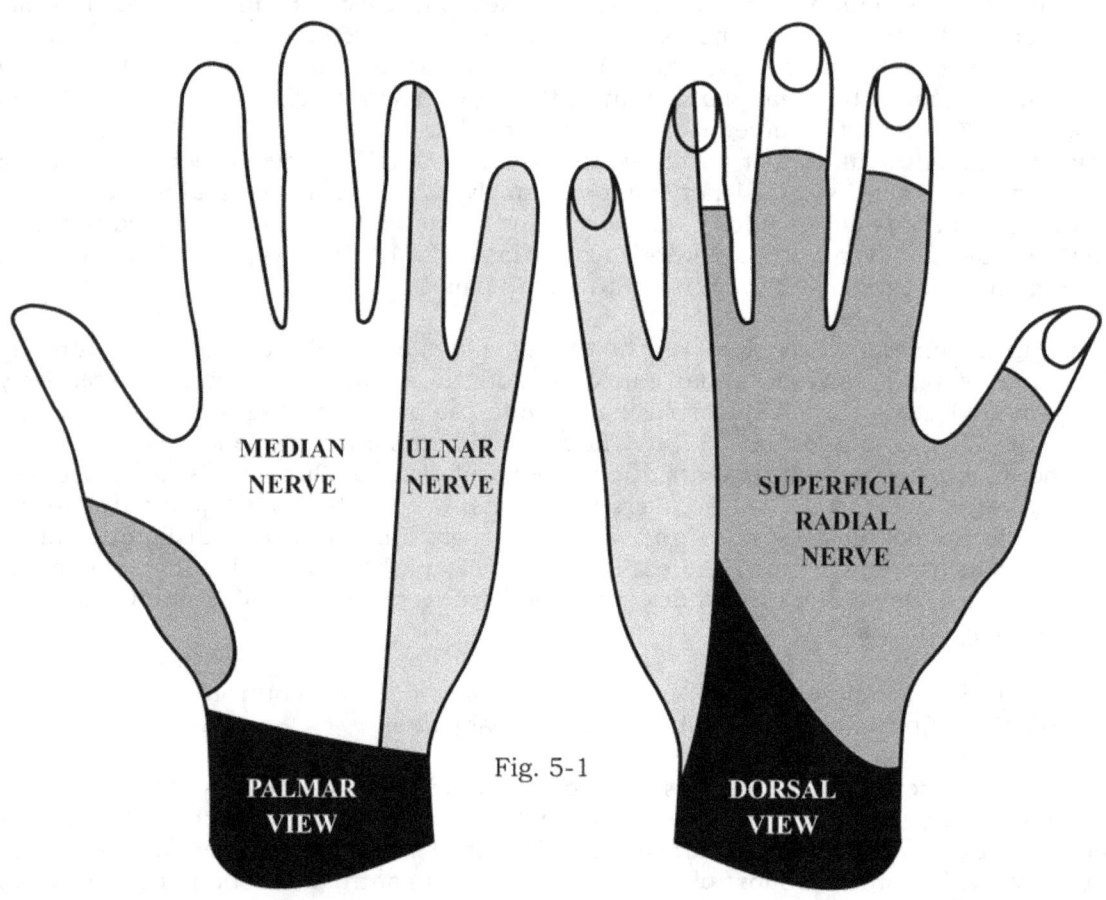

MEDIAN NERVE

ULNAR NERVE

SUPERFICIAL RADIAL NERVE

Fig. 5-1

PALMAR VIEW

DORSAL VIEW

Fig. 5-1: Cutaneous innervation of the hand. Median nerve fibers innervate white skin areas, ulnar nerve fibers innervate light gray areas, and superficial radial nerve fibers innervate medium gray areas.

19. What are the symptoms of carpal tunnel syndrome?

Entrapment or compression of the median nerve along its course through the carpal tunnel can produce median nerve damage from mechanical deformation and/or ischemia. In cases of insidious onset, pain and sensory deficits on the anterior surfaces of the lateral three and a half digits are typically the initial neurologic symptoms. Pain and hypalgesia may occur concomitantly (that is, simultaneously) as a result of the increased pressure both irritating some pain nerve fibers (thus producing pain) and damaging others (thereby diminishing sensitivity to pain). Although median nerve fibers provide cutaneous sensory innervation for the lateral two-thirds of the palm, pain and sensory deficits do not occur in the lateral two-thirds of the palm because the cutaneous median nerve fibers for the palm of the hand arise from the median nerve proximal to the carpal tunnel and extend across the wrist into the hand superficial to the flexor retinaculum. Motor deficits associated with dysfunction of the muscles of the thenar eminence occur subsequent to the onset of sensory deficits.

Conditions which produce such sensory and motor deficits are collectively referred to as the carpal tunnel syndrome. A broad spectrum of injuries, diseases, and conditions can produce carpal tunnel syndrome. Fractures and dislocations at the wrist, chronic inflammatory conditions such as rheumatoid arthritis and gout, and repetitive wrist movements (particularly those involving forceful flexion of the hand) can produce the syndrome by tenosynovitis of the flexor tendons in the carpal tunnel and the attendant thickening of the synovial sheaths. Systemic conditions such as hypothyroidism, diabetes mellitus, acromegaly, and systemic lupus erythematosus can produce the syndrome by retention of interstitial fluid and the attendant swelling of soft tissues.

20. Three spinal nerves (C6, C7, and C8) provide almost all the cutaneous sensory nerve fibers for the palmar surface of the hand and its thumb and fingers. The anterior surface at the tip of one of the hand's digits is part of the C6 dermatome, a second is part of the C7 dermatome, and a third is part of the C8 dermatome. In the drawing of Fig. 5-2, mark the digits whose tips are parts of the C6, C7, and C8 dermatomes.

21. Opponens pollicis is the sole prime mover of opposition of the thumb, which is the movement by which the tip of the thumb's palmar surface, starting from the anatomical position, is brought into contact with the tips of the palmar surfaces of the fingers. Opponens pollicis also makes it possible for the tip of a person's thumb to be brought into contact with a specific, small region on the surface of the palm. In the drawing of Fig. 5-3, mark the region on the surface of the palm which can be touched by the tip of the thumb if and only if opponens pollicis is fully functional.

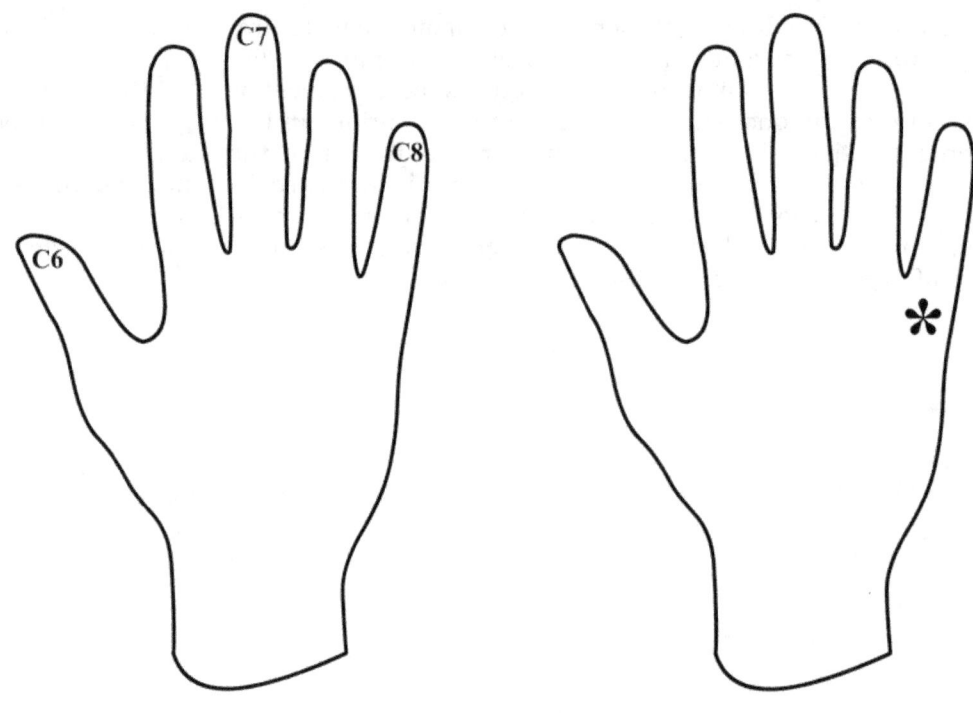

Fig. 5-2
Outline of the palmar surface
of the hand

Fig. 5-3
Outline of the palmar surface
of the hand

The asterisk in Fig. 5-3 marks the base of the little finger, which is the palmar region that overlies the head of the 5th metacarpal. A person can touch the base of the little finger with the tip of the thumb if and only if the thumb is fully functional.

22. During an examination of an injured person who has been delivered by ambulance to the ER, the physician inserts a piece of paper between the patient's index and middle fingers and asks the patient to resist the attempt by the physician to remove the piece of paper. The physician employs this test to assess the integrity of a major nerve of the upper limb. Identify the nerve and explain how this test assesses its integrity.

A person can grip a piece of paper between the index and middle fingers by simultaneously contracting the palmar interosseous muscle that acts on the index finger (which is the 1st palmar interosseous) and the dorsal interosseous muscle that acts on the lateral side of the middle finger (which is the 2nd dorsal interosseous). The 1st palmar interossus adducts the index finger against the middle finger, and the 2nd dorsal interosseous abducts the middle finger against the index finger; the combined actions press the index and middle fingers together. Both intrinsic hand muscles are innervated by the deep branch of the ulnar nerve. The test assesses the integrity of the ulnar nerve from its origin in the axilla to its division into its deep and superficial branches in the hand.

23. There are two arched arteries in the palm of the hand called the superficial and deep palmar arches. The superficial palmar arch lies anterior to the flexor digitorum tendons in the palm of the hand, and the deep palmar arch lies posterior to these tendons. In most persons, both the radial and ulnar arteries supply both arterial arches. Describe the Allen test and explain its purpose.

The Allen test is a physical test which can determine whether the ulnar and radial arteries are each a chief source of blood supply to a patient's hand. In the critical care setting, an arterial line is sometimes put into the radial artery near the wrist to monitor blood pressure and blood gases. The Allen test must be performed prior to placing the arterial line in the radial artery. The reason for the absolute necessity for performing the Allen test is because the ulnar and radial arteries are not always each a chief source of blood supply to the hand. If, for example, the ulnar artery cannot adequately supply a patient's hand in the absence of supply by the radial artery, and the patient's radial artery becomes blocked following the placement of an arterial line, the patient's hand would no longer receive adequate blood supply.

In the Allen test, each artery is tested separately as follows: The patient is asked first to make a tight fist. The making of a tight fist increases external pressure upon the capillary beds within the skin of the palm and thereby empties them of much of their blood content. The examiner next applies pressure to both the ulnar and radial arteries at their palpation sites in the anterior forearm. This significantly blocks the blood flow into the palmar arches. The patient is then asked to unclench the fist, exposing the blanched pallor of the palmar surface. Finally, the examiner releases the pressure on one artery and ascertains whether color is restored to the palmar surface within 2 to 4 seconds. The return of color indicates that the artery is a chief source of blood supply to the hand. The entire test is then repeated with the other artery to ascertain that both arteries are supplying the palmar arches.

24. What is the anatomical and physiological basis for the fact that, in a normal Allen test, a blood capillary bed in the palm of the hand that has been depleted of its blood supply should become filled with blood within a 2 to 4 second interval after blood supply has been fully restored?

The average velocity of blood flow in a capillary is 1 mm/sec, and the average length of a blood capillary is 1 mm. Consequently, a subcutaneous blood capillary bed emptied of much of its blood content should, under normal conditions, become filled with blood within about 2 seconds after blood supply is restored.

GLUTEAL REGION – Part A: Questions

Dissection of the gluteal region in gross lab focuses on identification of the 9 gluteal muscles, the sciatic nerve, and the superior and inferior gluteal nerves and arteries. Three topics dominate the gluteal anatomy most frequently applied in clinical practice: (1) palpation of the highest point of the iliac crest as an indicator of the level of the spinous process of the 4th lumbar vertebra, (2) hip fractures, and (3) the use of the buttock as a site for intramuscular injection.

Bones & Joints of the Gluteal Region

The two most inferior bones of the adult spine (the sacrum and coccyx), the coxal bone, and the proximal, or upper, end of the femur form the skeletal framework of the gluteal region. We will refer to the hip bone as the coxal bone in this workbook because orthopedists use the term hip to refer to the proximal end of the femur. The hip consists of the head of the femur, the neck of the femur, and the greater and lesser trochanters of the femur. The sacrum, coxal bone, and hip form two synovial joints in the gluteal region: the sacroiliac and hip joints.

1. Which of the following parts of the coxal bone is (are) palpable in the lower trunk and gluteal regions of the body?
   _____ Anterior inferior iliac spine   _____ Iliac crest   _____ Pubic tubercle
   _____ Anterior superior iliac spine   _____ Ischial tuberosity
   _____ Posterior inferior iliac spine   _____ Ischial spine
   _____ Posterior superior iliac spine   _____ Pubic symphysis

2. Which of the following parts of the proximal femur is (are) palpable in the gluteal region?
   _____ Greater trochanter   _____Lesser trochanter   _____ Neck of the femur

3. Explain why most displaced, intracapsular hip fractures in elderly individuals lead to avascular necrosis of the head of the femur (that is, death of the head of the femur because of loss of blood supply).

4. What are pelvic fractures, and why do they commonly occur as a pair of fractures or a fracture accompanied by a dislocated joint?

5. Explain why a person suffering from a painful hip joint effusion is most comfortable seated with the painful thigh slightly abducted and externally rotated at the hip joint.

Muscles of the Gluteal Region

The 9 muscles of the gluteal region are listed here from A to I. Refer to this list in answering questions 6-8.
   A. Gluteus maximus   D. Tensor fasciae latae   G. Inferior gemellus
   B. Gluteus medius   E. Piriformis   H. Obturator internus
   C. Gluteus minimus   F. Superior gemellus   I. Quadratus femoris

6. _____, _____, _____, _____, and _____ can externally rotate the thigh at the hip joint.

7. _____, _____, and _____ can abduct the thigh at the hip joint.

8. _____ can extend the thigh at the hip joint.

Identify the nerve or nerve fibers which innervate each gluteal muscle:

9. Gluteus maximus: _____

10. Gluteus medius: _____

11. Gluteus minimus: _____

12. Tensor fasciae latae: _____

13. Piriformis: _____

14. Superior gemellus: _____

15. Inferior gemellus: _____

16. Obturator internus: _____

17. Quadratus femoris: _____

18. Gluteus maximus exerts major roles in the walking gait. Explain the roles exerted by gluteus maximus during (a) the terminal swing phase (TSW) of the swing period and (b) the initial contact (IC) and loading response (LR) phases of the following stance period.

19. Explain the lateral pelvic tilting action exerted by gluteus medius and gluteus minimus during the initial contact (IC), loading response (LR), and mid stance (MST) phases of the stance period.

20. In the pelvis on each side, nerve fibers from the anterior rami of several spinal nerves form a plexus, or network, of nerve fibers called the sacral plexus. The sacral plexus gives rise to all the nerves that innervate the gluteal muscles. Which spinal nerves contribute nerve fibers to the sacral plexus?
    _____ L1   _____ S1
    _____ L2   _____ S2
    _____ L3   _____ S3
    _____ L4   _____ S4
    _____ L5   _____ S5

21. The sacral plexus gives rise to the sciatic nerve, which is the largest nerve in the body. The sciatic nerve is derived from nerve fibers in the anterior rami of which spinal nerves?
    _____ L1   _____ L2   _____ L3   _____ L4   _____ L5
    _____ S1   _____ S2   _____ S3   _____ S4   _____ S5

22. The sciatic nerve extends from the sacral plexus into the gluteal region by passing through the greater sciatic foramen. Describe the course of the sciatic nerve through the gluteal region, and explain why the upper lateral quadrant of the buttock is the safest quadrant for intramuscular injections.

23. What is the clinical importance of the anatomical fact that the highest point of the iliac crest marks the level of the spinous process of the 4th lumbar vertebra in the spine?

END OF QUESTIONS IN PART A OF THE CHAPTER ON THE GLUTEAL REGION

GLUTEAL REGION – Part B: Questions and Answers

Dissection of the gluteal region in gross lab focuses on identification of the 9 gluteal muscles, the sciatic nerve, and the superior and inferior gluteal nerves and arteries. Three topics dominate the gluteal anatomy most frequently applied in clinical practice: (1) palpation of the highest point of the iliac crest as an indicator of the level of the spinous process of the 4th lumbar vertebra, (2) hip fractures, and (3) the use of the buttock as a site for intramuscular injection.

The Walking Gait

The questions on the bones, joints, and muscles of the lower limb in the following chapters will focus on how these musculoskeletal structures function to bear the weight of the upper body and to provide for the walking gait. We therefore begin with a general brief description of the human walking gait.

The major features of the walking gait involve movements at the hip, knee, ankle, and metatarsophalangeal joints. The metatarsophalangeal joints are the joints in the foot between the metatarsals and the proximal phalanges of the toes.

A complete cycle of the walking gait is called a stride. Each lower limb passes through two periods during a stride: a stance period, during which the foot is in contact with the surface below, and a swing period, during which the foot is swung forward above the surface below.

A stride is said to begin with the beginning of the stance period. The first moment of the stance period is called heel strike because, after the lower limb has been swung forward, the heel of the foot strikes the surface below to begin the stance period (Fig. 6-1). Heel strike begins the first phase of the stance period, which is called the initial contact (IC) phase. The initial contact phase is of very short duration and positions the lower limb to accept support of upper body weight.

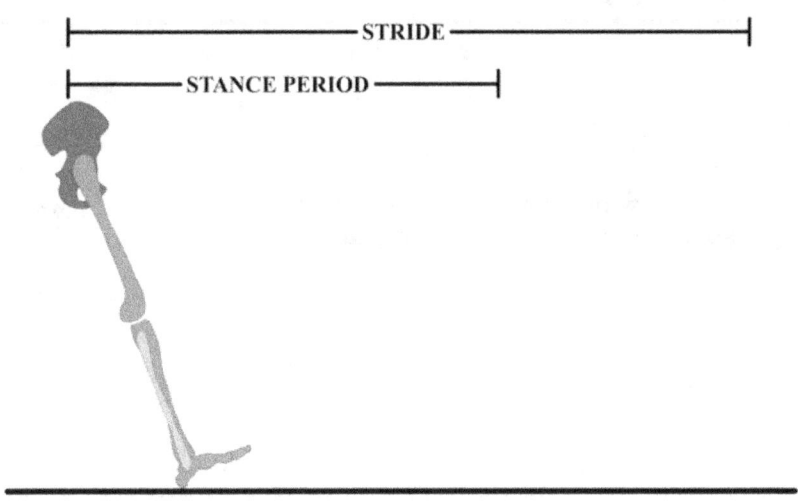

Fig. 6-1: Position of the lower limb at heel strike.

The next phase of the stance period is called the loading response (LR) phase because the lower limb responds to the loading of upper body weight upon it (Fig. 6-2). The lower limb rolls forward on its heel during the loading response phase to help sustain the forward momentum of the body. This mechanism by which the body rocks forward over the heel of the newly planted foot is called heel rocker action.

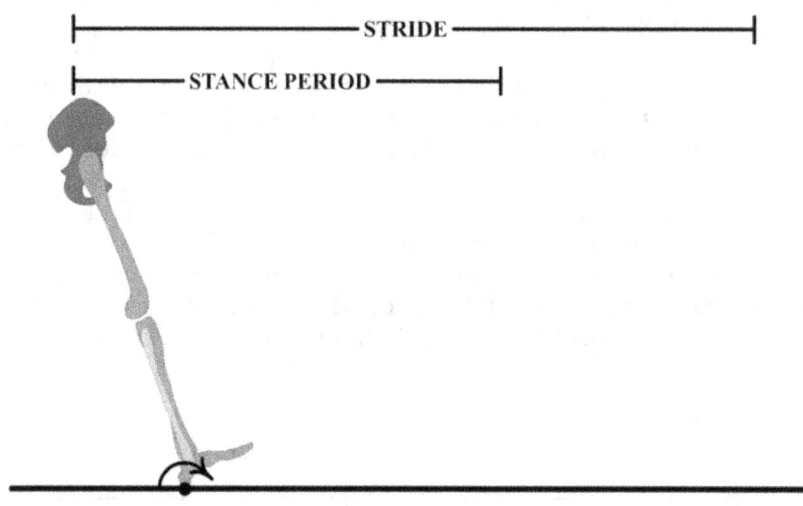

Fig. 6-2: Curved arrow showing how the lower limb rolls, or rocks, forward around its heel during the loading response phase of the walking gait.

Heel rocker action brings the foot into full contact with the surface below, and the moment at which full contact occurs is called foot flat (Fig. 6-3). Foot flat marks the end of the loading response phase and the beginning of the third phase of the stance period, the mid stance phase.

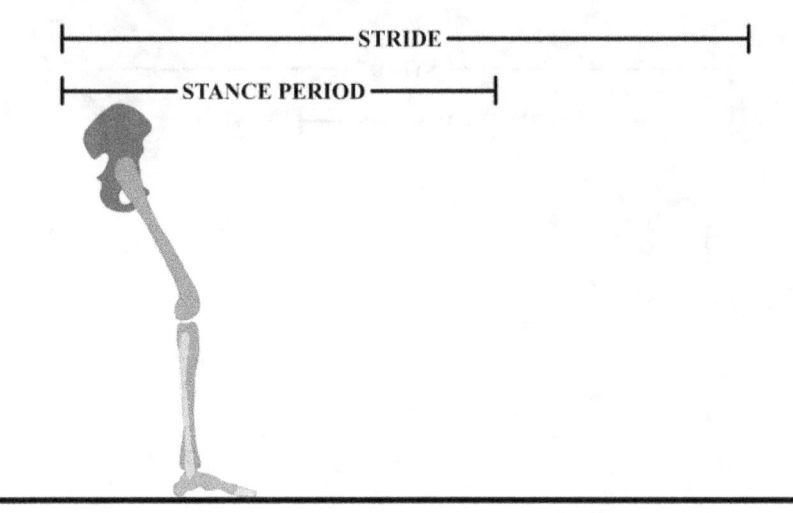

Fig. 6-3: Position of the lower limb at foot flat.

The third phase of the stance period is called the mid stance (MST) phase because it occupies the mid time phase of the stance period. An ankle rocker action (in which the lower limb rolls forward around the ankle joint) helps sustain forward body momentum in the mid stance phase (Fig. 6-4). Ankle rocker action helps bring the upper body weight directly over the fully planted foot.

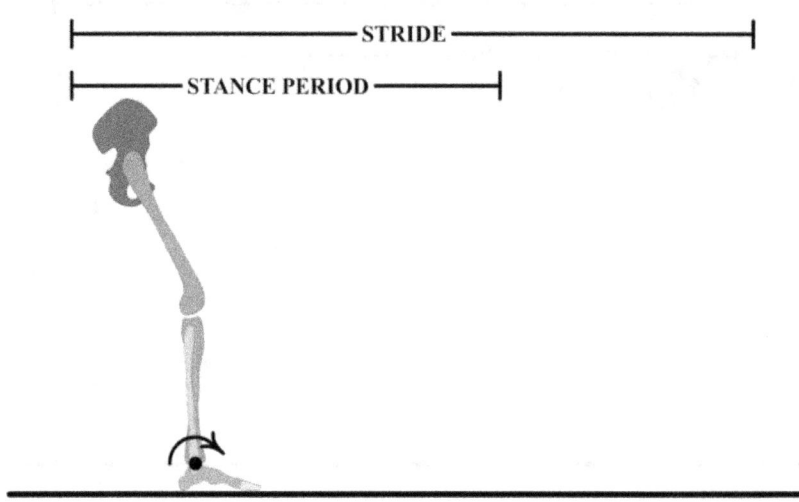

Fig. 6-4: Curved arrow showing how the lower limb rolls, or rocks, forward around its ankle joint during the mid stance phase of the walking gait.

The fourth phase of the stance period is called the terminal stance (TST) phase. This phase begins with the raising of the heel, called either heel rise or heel off (Fig. 6-5). A forefoot rocker action (in which the lower limb rolls forward around the metatarsophalangeal joints) helps sustain the forward body momentum. Forefoot rocker action helps draw the heel up and advance the body ahead of the lower limb.

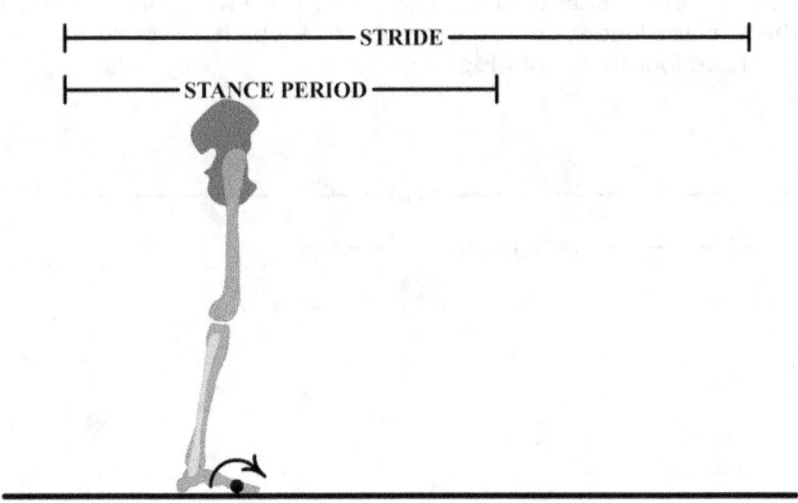

Fig. 6-5: Curved arrow showing how the lower limb starts to roll, or rock, forward around its metatarsophalangeal joints at heel rise, the moment at which the terminal stance phase of the walking gait begins.

The fifth and final phase of the stance period is called the pre swing (PSW) phase. The pre swing phase is so named because the lower limb becomes positioned to swing forward rapidly beneath the advancing upper body. Upper body weight is unloaded from the lower limb and transferred to the contralateral lower limb during the pre swing phase (the contralateral lower limb is the limb on the other side of the body). The pre swing phase ends with the forefoot rolling off from the surface below, the final moment of which is called toe off because the big toe is the last part of the foot to roll off (Fig. 6-6).

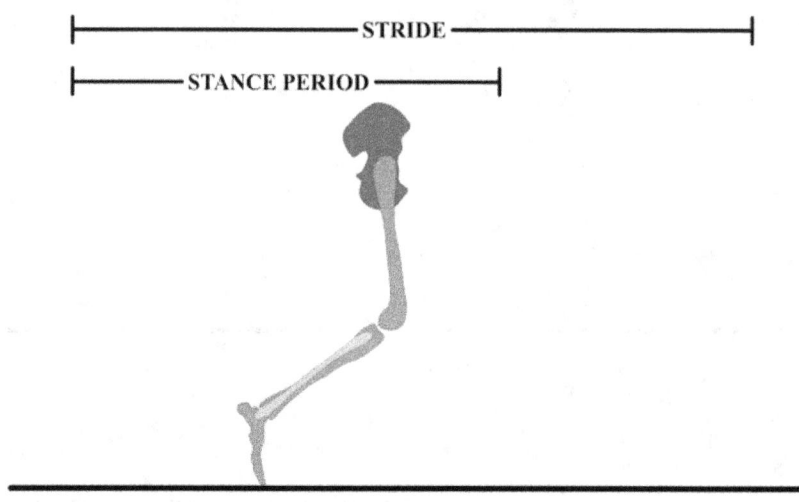

Fig. 6-6: Position of the lower limb at the end of the pre swing phase. In a moment, the big toe will roll off the surface below to begin the first phase of the swing period, which is called the initial swing phase.

The first phase of the swing period is called the initial swing (ISW) phase. The foot of the lower limb is lifted above the surface below and the entire limb is accelerated forward. This forward acceleration of the lower limb helps provide the force to sustain forward body momentum. The mid phase of the swing period is called the mid swing (MSW) phase. During the mid swing phase, the lower limb passes beneath the upper body to a position in which the tibia of the leg is vertical (Fig. 6-7). The third and final phase of the swing period is called the terminal swing (TSW) phase. The lower limb decelerates its forward momentum during the terminal swing phase in preparation for the initial contact phase of the next gait cycle.

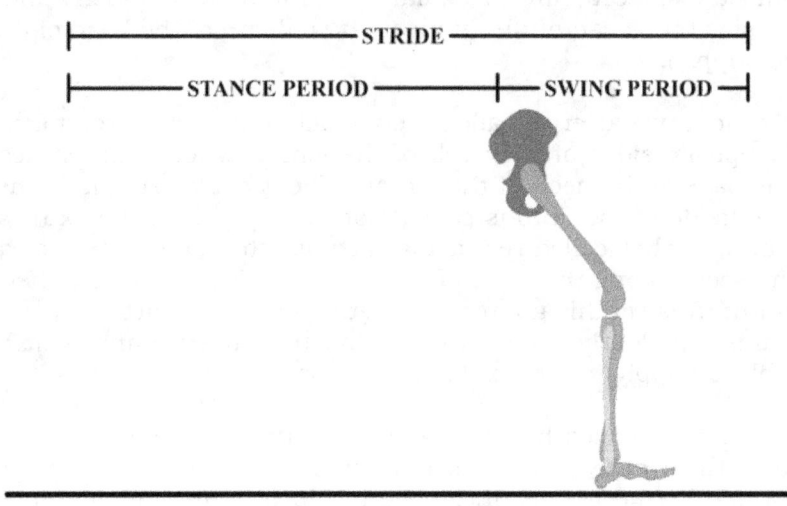

Fig. 6-7: Position of the lower limb when the tibia of the leg is vertical during the mid swing phase of the walking gait.

Bones & Joints of the Gluteal Region

The two most inferior bones of the adult spine (the sacrum and coccyx), the coxal bone, and the proximal, or upper, end of the femur form the skeletal framework of the gluteal region. We will refer to the hip bone as the coxal bone in this workbook because orthopedists use the term hip to refer to the proximal end of the femur. The hip consists of the head of the femur, the neck of the femur, and the greater and lesser trochanters of the femur. The sacrum, coxal bone, and hip form two synovial joints in the gluteal region: the sacroiliac and hip joints.

1. Which of the following parts of the coxal bone is (are) palpable in the lower trunk and gluteal regions of the body?

    _____ Anterior inferior iliac spine   __x__ Iliac crest   __x__ Pubic tubercle
    __x__ Anterior superior iliac spine   __x__ Ischial tuberosity
    _____ Posterior inferior iliac spine _____ Ischial spine
    __x__ Posterior superior iliac spine __x__ Pubic symphysis

2. Which of the following parts of the proximal femur is (are) palpable in the gluteal region?

    __x__ Greater trochanter   _____ Lesser trochanter   _____ Neck of the femur

3. Explain why most displaced, intracapsular hip fractures in elderly individuals lead to avascular necrosis of the head of the femur (that is, death of the head of the femur because of loss of blood supply).

In the gluteal region, two arteries called the medial and lateral circumflex femoral arteries curve around opposite sides of the neck of the femur to form an extracapsular vascular ring around the base of the neck of the femur. The vascular ring is termed extracapsular because it lies outside of the fibrous capsule of the hip joint. The extracapsular vascular ring gives rise to branches called retinacular arteries that penetrate the capsule of the hip joint and then ascend along the neck of the femur until they enter either the neck or the head of the femur. It is via this approach that the retinacular arteries supply the neck and head of the femur. In elderly persons, the retinacular arteries and their branches are the sole source of blood supply to the head of the femur.

Fractures of the hip are called hip fractures. Fractures of the femoral neck are especially common among the elderly; the contribution of osteoporosis to the genesis of such fractures accounts for the greater incidence of hip fractures among elderly women than among elderly men. Most femoral neck fractures in elderly individuals occur immediately distal to the head of the femur and thus are called subcapital fractures. Subcapital fractures are intracapsular fractures; that is, they lie deep to the fibrous capsule of the hip joint and are surrounded by the joint's synovial cavity. Many intracapsular fractures are also displaced fractures, which means that the two bone fragments at the fracture site have been moved out of normal alignment.

Most displaced, intracapsular fractures of the neck of the femur typically rupture the retinacular arteries and thus significantly disrupt femoral head blood supply. Moreover. deposition of new bone tissue at intracapsular fracture sites is hindered by the relatively small amount of soft tissues around the sites; this is because the soft tissues that surround a bone fracture play an important role in the deposition of new bone tissue at the fracture site. The combination of these circumstances increases the risk that the damaged or torn retinacular arteries will not become sufficiently repaired to restore adequate blood supply to the head of the femur. Extracapsular femoral neck fractures generally do not disrupt femoral head blood supply, and thus are associated with a lower incidence of avascular necrosis of the femoral head.

4. What are pelvic fractures, and why do they commonly occur as a pair of fractures or a fracture accompanied by a dislocated joint?

In the lower part of the trunk of the body, the left and right coxal bones are joined with each other and the two lowest bones of the spine (the sacrum and coccyx) to form a bowl-shaped ring of bones called the bony pelvis. In the front of the bony pelvis, a cartilaginous joint called the pubic symphysis joins the pubic parts of the left and right coxal bones. In the back of the bony pelvis, a synovial joint called the sacroiliac joint joins the iliac part of the coxal bone on each side with the sacrum. In the back of the bony pelvis, a cartilaginous joint called the sacrococcygeal joint joins the sacrum and coccyx. The joints of the bony pelvis permit very little movement of the bones within the bony pelvis.

Fractures of the bony pelvis are called pelvic fractures. Since the bony pelvis is structurally a united ring of bones, breaks within the ring generally occur in pairs. The pair of breaks commonly is either a pair of fractures or a fracture accompanied by a joint dislocation. The superior and inferior pubic rami are the most commonly fractured parts of the bony pelvis; dislocation of the pubic symphysis is more common than dislocation of the sacroiliac joints. There is high morbidity and mortality associated with pelvic fractures because of attendant hemorrhagic shock and pelvic organ damage. In particular, it must always be assumed with pelvic fractures, until proven otherwise, that the bladder and urethra are also damaged.

5. Explain why a person suffering from a painful hip joint effusion is most comfortable seated with the painful thigh slightly abducted and externally rotated at the hip joint.

Injury or infection of a synovial joint typically leads to increased production of synovial fluid within the joint. The increased volume of synovial fluid within the joint is called a joint effusion. Joint effusions become uncomfortable or painful as tension is increased on the synovial membrane lining the joint. A person can reduce the tension on the synovial membrane of an effused hip joint by sitting with the thigh slightly abducted and externally rotated at the hip joint. This position reduces the tension to a minimum because it maximizes the encapsulation of the femoral head by the acetabular cavity and labrum.

Muscles of the Gluteal Region

The 9 muscles of the gluteal region are listed here from A to I. Refer to this list in answering questions 6-8.

A. Gluteus maximus   D. Tensor fasciae latae   G. Inferior gemellus
B. Gluteus medius   E. Piriformis   H. Obturator internus
C. Gluteus minimus   F. Superior gemellus   I. Quadratus femoris

6. __E__, __F__, __G__, __H__, and __I__ can externally rotate the thigh at the hip joint.

7. __B__, __C__, and __D__ can abduct the thigh at the hip joint.

8. __A__ can extend the thigh at the hip joint.

Identify the nerve or nerve fibers which innervate each gluteal muscle:

9. Gluteus maximus: __inferior gluteal nerve__

10. Gluteus medius: __superior gluteal nerve__

11. Gluteus minimus: __superior gluteal nerve__

12. Tensor fasciae latae: __superior gluteal nerve__

13. Piriformis: __nerve fibers from S1 and S2__

14. Superior gemellus: __nerve to obturator internus__

15. Inferior gemellus: __nerve to quadratus femoris__

16. Obturator internus: __nerve to obturator internus__

17. Quadratus femoris: __nerve to quadratus femoris__

18. Gluteus maximus exerts major roles in the walking gait. Explain the roles exerted by gluteus maximus during (a) the terminal swing phase (TSW) of the swing period and (b) the initial contact (IC) and loading response (LR) phases of the following stance period.

This is the first of several questions in this workbook on the roles of lower limb muscles in the walking gait. It is important to recognize the following two general rules regarding lower limb muscle actions in the walking gait: If the lower limb is in the stance period, lower limb muscles act to accelerate, stabilize, or decelerate the movement of their origin, or proximal attachment site. If, on the other hand, the lower limb is in the swing period, lower limb muscles act to accelerate, stabilize, or decelerate the movement of their insertion, or distal attachment site. Gluteus maximus' actions in the walking gait exemplify these two general rules.

During the TSW phase of the swing period, when the lower limb is being swung forward at the hip joint in anticipation of heel strike, gluteus maximus acts to decelerate the movement of the part of the lower limb to which it is attached distally, namely, the thigh. Gluteus maximus acts during the late part of the TSW phase to restrain and slow down forward movement of the thigh by pulling on the back of the thigh from the bony pelvis.

During the IC and LR phases of the stance period, gluteus maximus acts to stabilize the part of the lower limb to which it is attached proximally, namely, the bony pelvis. The trunk of the body has a tendency to fall forward at the hip joint during the IC and LR phases because upper body weight is being loaded upon the limb along a line that projects anterior to the hip joint. The projection of the upper body weight line anterior to the hip joint generates a significant flexion torque at the joint that pulls the body trunk forward at the joint. Gluteus maximus opposes this flexion torque by pulling on the back of the bony pelvis from the thigh. Individuals who suffer from paralysis of gluteus maximus adopt an abnormal gait called the gluteus maximus gait in which they lean the body trunk backward at heel strike in order to compensate for the loss of the muscle's contribution to the IC and LR phases of the walking gait.

19. Explain the lateral pelvic tilting action exerted by gluteus medius and gluteus minimus during the initial contact (IC), loading response (LR), and mid stance (MST) phases of the stance period.

 During the IC, LR and MST phases, gluteus medius and minimus act to stabilize the part of the lower limb to which they are attached proximally, namely, the bony pelvis. Gluteus medius and minimus oppose the tendency of the upper body during these three phases to fall medially downward at the hip joint (to fall downward on the medial side of the hip joint. Let's examine this action by gluteus medius and minimus as the right lower limb passes through the IC, LR, and MST phases. As the right lower limb passes through the IC and LR phases of the stance period, upper body weight is loaded upon the right lower limb at the same time that upper body weight is unloaded from the left lower limb. In other words, left lower limb support of upper body weight decreases as the right lower limb goes through the IC and LR phases; this loss of left lower limb support causes the left side of the pelvis to drop, or sag. In other words, the pelvis tilts downward, or medially, at the right hip joint. The gluteus medius and minimus muscles around the right hip joint, however, minimize this downward, or medial, tilt of the pelvis at the right hip joint by pulling on the bony pelvis from their attachment to the femur. Because this pull tilts the pelvis upward, or laterally, at the right hip joint, the pull is called a lateral pelvic tilting action (Fig. 6-8). When walking, gluteus medius and minimus are the chief muscles responsible for the lateral pelvic tilting action exerted during the IC, LR, and MST phases. The bony pelvis typically tilts 3° to 4° downward at a lower limb's hip joint as the lower limb passes through the MST phase.

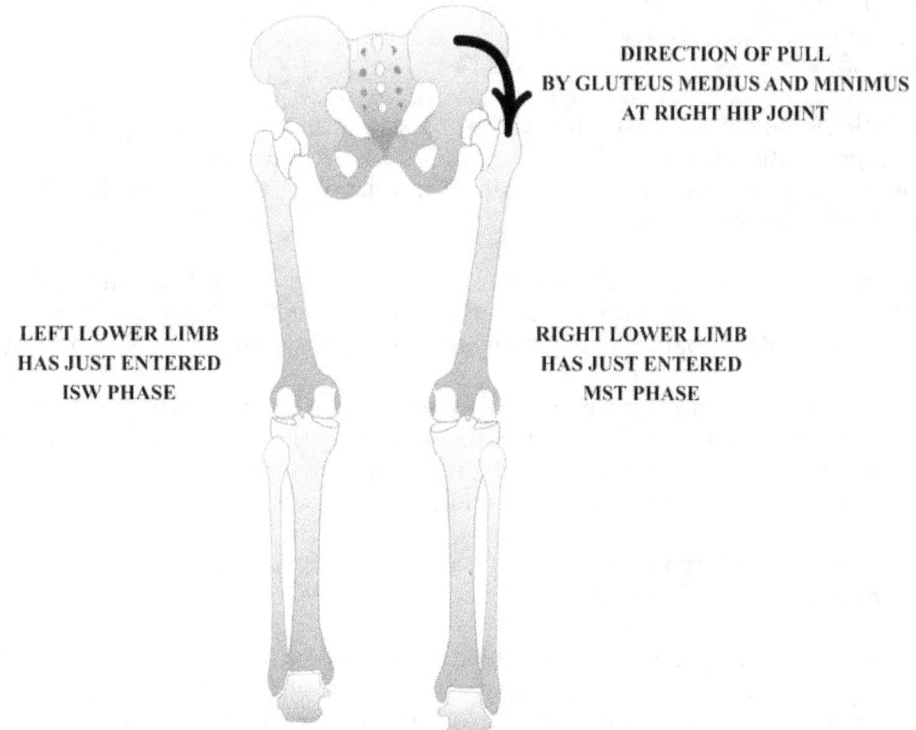

DIRECTION OF PULL
BY GLUTEUS MEDIUS AND MINIMUS
AT RIGHT HIP JOINT

LEFT LOWER LIMB
HAS JUST ENTERED
ISW PHASE

RIGHT LOWER LIMB
HAS JUST ENTERED
MST PHASE

Fig. 6-8: Posterior view of lower limbs at the moment in the walking gait when the left lower limb has just entered the initial swing phase of the swing period and the right lower limb has just entered the mid stance phase of the stance period. As the curved arrow indicates, gluteus medius and minimus on the right side are pulling upward, or laterally, on the bony pelvis to minimize the downward tilt of the bony pelvis at the right hip joint to just 3 to 4°.

Individuals who suffer from weakness or paralysis of gluteus medius and minimus compensate for the loss of the lateral pelvic tilting action of these muscles by adopting an abnormal gait called the gluteus medius gait. To examine this gait, consider a person whose gluteus medius and minimus muscles are weak in the right lower limb. When this person walks, he/she pushes off with the left foot (as the left lower limb passes through the pre swing phase of the stance period) to lean the upper body toward the right and temporarily balance it over the right lower limb as the right lower limb passes through the LR, MST, and TST phases.

20. In the pelvis on each side, nerve fibers from the anterior rami of several spinal nerves form a plexus, or network, of nerve fibers called the sacral plexus. The sacral plexus gives rise to all the nerves that innervate the gluteal muscles. Which spinal nerves contribute nerve fibers to the sacral plexus?

_____ L1	__x__ S1
_____ L2	__x__ S2
_____ L3	__x__ S3
__x__ L4	__x__ S4
__x__ L5	_____ S5

21. The sacral plexus gives rise to the sciatic nerve, which is the largest nerve in the body. The sciatic nerve is derived from nerve fibers in the anterior rami of which spinal nerves?

_____ L1	__x__ S1
_____ L2	__x__ S2
_____ L3	__x__ S3
__x__ L4	_____ S4
__x__ L5	_____ S5

22. The sciatic nerve extends from the sacral plexus into the gluteal region by passing through the greater sciatic foramen. Describe the course of the sciatic nerve through the gluteal region, and explain why the upper lateral quadrant of the buttock is the safest quadrant for intramuscular injections.

The sciatic nerve follows an inferolateral course through the lower medial quadrant of the gluteal region before extending inferiorly into the back of the thigh at a point midway between the ischial tuberosity of the coxal bone and the greater trochanter of the femur.

The sciatic nerve is the single most important gluteal structure at risk of injury by an intramuscular injection in the buttock. Because the sciatic nerve passes through the lower medial quadrant of the buttock, the upper lateral quadrant of the buttock is the safest quadrant for intramuscular injections.

23. What is the clinical importance of the anatomical fact that the highest point of the iliac crest marks the level of the spinous process of the 4th lumbar vertebra in the spine?

A spinal tap is the procedure by which a small volume of cerebrospinal fluid (CSF) is withdrawn from the subarachnoid space that surrounds the spinal cord and the roots of the spinal nerves in the spine. In an adult, the lower end of the spinal cord, which is called the conus medullaris, lies at the level of the lower border of the body of the 1st lumbar vertebra. Consequently, if a needle is inserted into the spine between the spinous processes of the 3rd and 4th or 4th and 5th lumbar vertebrae to collect CSF, there is virtually no possibility that the needle will accidentally impinge upon the spinal cord (however, there is some risk that a spinal nerve root may be grazed by the needle). The highest point of the iliac crest is a highly reliable landmark for the level of the spinous process of the 4th lumbar vertebra in the spine.

Dissection of the thigh and knee in gross lab focuses on identification of the 16 thigh muscles, the three major nerves of the thigh (the sciatic, femoral, and obturator nerves), the courses of the femoral artery and its major branches through the thigh, the courses of the femoral and great saphenous veins through the thigh, the courses of the tibial nerve, common fibular nerve, popliteal artery, and popliteal vein across the knee, and the menisci and major ligaments of the knee joint. Five topics dominate the thigh and knee anatomy most frequently applied in clinical practice: (1) the relationships among the femoral nerve, artery and vein in the uppermost part of the femoral triangle, (2) the nature of the three knee injuries which commonly occur when the knee is hit by a very powerful valgus force, (3) the innervation and major actions of thigh muscles, (4) the spinal cord levels at which spinal cord reflexes are assessed by the quadriceps femoris deep tendon reflex, and (5) the location and drainage areas of the lymph nodes in the back of the knee and the superficial lymph node groups clustered in the upper anterior region of the thigh.

Bones & Joint Associated with the Actions of Thigh Muscles

The 16 muscles of the thigh move the thigh at the hip joint and the leg at the knee joint. The knee joint provides for flexion and extension of the leg. If the leg is flexed at least 30°, the knee joint can also provide for internal and external rotation of the leg. In the knee joint, the patella articulates with the patellar surface of the femur, the medial femoral condyle articulates with both the medial meniscus and medial tibial condyle, and the lateral femoral condyle articulates with both the lateral meniscus and lateral tibial condyle.

1. The prominent bony bumps that can be palpated on the medial and lateral sides of the distal end of the femur are the _____.

2. The prominent bony bumps that can be palpated on the medial and lateral sides of the proximal end of the tibia are the _____.

3. The prominent bony bump that can be palpated on the anterior surface of the tibia about an inch below the tibial condyles is the _____.

4. The prominent bony bump that can be palpated at the upper end of the fibula is the

   _____.

5. Which three knee injuries commonly occur when the knee is hit by a very powerful valgus force?

6. Describe the locations of the 4 bursae around the anterior aspect of the knee. Which of the bursae communicate with the synovial cavity of the knee joint?

7. What is the bulge sign?

8. Describe the locations of the 3 bursae around the posterior aspect of the knee. Which of the bursae communicate with the synovial cavity of the knee joint?

9. What is a Baker's cyst?

10. Describe the lower limb tissues whose lymph is filtered by the lymph nodes in the back of the knee.

The 16 muscles of the thigh are listed here from A to R. Adductor magnus has two parts; each part is listed separately. Biceps femoris has two heads of origin; each head is listed separately. Refer to this list in answering questions 11-19.

A. Adductor brevis
B. Adductor longus
C. Adductor magnus
   (the adductor part)
D. Adductor magnus
   (the hamstring part)
E. Biceps femoris (the long head)
F. Biceps femoris (the short head)

G. Gracilis
H. Iliacus
I. Obturator externus
J. Pectineus
K. Psoas major
L. Rectus femoris

M. Sartorius
N. Semimembranosus
O. Semitendinosus
P. Vastus intermedius
Q. Vastus lateralis
R. Vastus medialis

11. _____, _____, and _____ are the only thigh muscles which can both extend the thigh at the hip joint and flex the leg at the knee joint; these three muscles are all posterior thigh muscles and are commonly referred to as the hamstrings.

12. _____, _____, _____, and _____ are the only thigh muscles which can extend the leg at the knee joint. These four muscles are all anterior thigh muscles and are commonly referred to as the quadriceps femoris. They all share a common tendon of insertion in which the patella is embedded.

13. _____ and _____ are the two anterior thigh muscles that act together as the most powerful flexors of the thigh at the hip joint. They are commonly referred to as iliopsoas.

14. _____, _____, _____, _____, and _____ are the 5 thigh muscles which can adduct the thigh at the hip joint.

15. _____ and _____ are the only thigh muscles which can externally rotate the leg at the knee joint.

16. _____, _____, _____, and _____ are the only thigh muscles which can internally rotate the leg at the knee joint.

17. _____ is the only medial thigh muscle which can extend the thigh at the hip joint.

18. _____ can both flex and adduct the thigh at the hip joint.

19. _____ can externally rotate the thigh at the hip joint.

The nerves and nerve fibers which innervate the thigh muscles are listed here from A to E. Refer to this list in matching each thigh muscle with its innervation.

A. Femoral nerve
B. Obturator nerve
C. Common fibular portion of the sciatic nerve
D. Tibial portion of the sciatic nerve
E. Nerve fibers from the anterior rami of L1, L2, and L3

20. _____ Adductor brevis

21. _____ Adductor longus

22. _____ Adductor part of adductor magnus

23. _____ Hamstring part of adductor magnus

24. _____ Long head of biceps femoris

25. _____ Short head of biceps femoris

26. _____ Gracilis

27. _____ Iliacus

28. _____ Obturator externus

29. _____ and _____ Pectineus

30. _____ Psoas major

31. _____ Rectus femoris

32. _____ Sartorius

33. _____ Semimembranosus

34. _____ Semitendinosus

35. _____ Vastus intermedius

36. _____ Vastus lateralis

37. _____ Vastus medialis

38. In the abdomen on each side, nerve fibers from the anterior rami of several spinal nerves form a plexus, or network, of nerve fibers called the lumbar plexus. Which spinal nerves contribute nerve fibers to the lumbar plexus?

_____ L1	_____ S1
_____ L2	_____ S2
_____ L3	_____ S3
_____ L4	_____ S4
_____ L5	_____ S5

39. The lumbar plexus gives rise to the femoral nerve. Which spinal nerves contribute nerve fibers to the femoral nerve?

_____ L1	_____ S1
_____ L2	_____ S2
_____ L3	_____ S3
_____ L4	_____ S4
_____ L5	_____ S5

40. The lumbar plexus gives rise to the obturator nerve. Which spinal nerves contribute nerve fibers to the obturator nerve?

_____ L1	_____ S1
_____ L2	_____ S2
_____ L3	_____ S3
_____ L4	_____ S4
_____ L5	_____ S5

41. Explain why the quadriceps femoris tendon reflex test assesses spinal cord reflexes at the L2, L3, and L4 spinal cord levels.

42. Explain why it can be said that L1 and L2 control flexion of the thigh at the hip joint.

43. Explain why it can be said that L3 and L4 control extension of the leg at the knee joint.

44. The hamstrings (semimembranosus, semitendinosus, and the long head of biceps femoris) exert major roles in the walking gait. Explain the roles exerted by the hamstrings during (a) the terminal swing phase (TSW) of the swing period and (b) the initial contact (IC) and loading response (LR) phases of the stance period.

45. The hamstring part of adductor magnus exerts major roles in the walking gait. Explain the roles exerted by the hamstring part of adductor magnus during (a) the terminal swing phase (TSW) of the swing period and (b) the initial contact (IC) and loading response (LR) phases of the stance period.

46. The quadriceps femoris muscles exert major roles in the walking gait. Explain the roles exerted by the quadriceps femoris muscles during (a) the terminal swing phase (TSW) of the swing period and (b) the loading response (LR) phases of the stance period.

47. Adductor longus and gracilis act in unison to exert a major role during the early part of the initial swing phase (ISW) of the swing period. Explain this action.

48. Proceeding from lateral to medial, what are the relationships among the femoral nerve (N), femoral artery (A), femoral vein (V), and femoral canal (C) immediately below the inguinal ligament?

_____ NAVC	_____ ANVC	_____ VNAC	_____ CNAV
_____ NACV	_____ ANCV	_____ VNCA	_____ CNVA
_____ NVAC	_____ AVNC	_____ VANC	_____ CANV
_____ NVCA	_____ AVCN	_____ VACN	_____ CAVN
_____ NCAV	_____ ACNV	_____ VCAN	_____ CVAN
_____ NCVA	_____ ACVN	_____ VCNA	_____ CVNA

49. Where can the pulsations of the femoral artery be palpated immediately below the inguinal ligament?

50. Describe the lower limb tissues whose lymph is filtered by the vertical group of superficial inguinal lymph nodes.

51. Describe the lower limb tissues whose lymph is filtered by the horizontal group of superficial inguinal lymph nodes.

END OF QUESTIONS IN PART A OF THE CHAPTER ON THE THIGH AND KNEE

THIGH AND KNEE – Part B: Questions and Answers

Dissection of the thigh and knee in gross lab focuses on identification of the 16 thigh muscles, the three major nerves of the thigh (the sciatic, femoral, and obturator nerves), the courses of the femoral artery and its major branches through the thigh, the courses of the femoral and great saphenous veins through the thigh, the courses of the tibial nerve, common fibular nerve, popliteal artery, and popliteal vein across the knee, and the menisci and major ligaments of the knee joint.   Five topics dominate the thigh and knee anatomy most frequently applied in clinical practice: (1) the relationships among the femoral nerve, artery and vein in the uppermost part of the femoral triangle, (2) the nature of the three knee injuries which commonly occur when the knee is hit by a very powerful valgus force, (3) the innervation and major actions of thigh muscles, (4) the spinal cord levels at which spinal cord reflexes are assessed by the quadriceps femoris deep tendon reflex, and (5) the location and drainage areas of the lymph nodes in the back of the knee and the superficial lymph node groups clustered in the upper anterior region of the thigh.

Bones & Joint Associated with the Actions of Thigh Muscles

The 16 muscles of the thigh move the thigh at the hip joint and the leg at the knee joint. The knee joint provides for flexion and extension of the leg. If the leg is flexed at least 30°, the knee joint can also provide for internal and external rotation of the leg. In the knee joint, the patella articulates with the patellar surface of the femur, the medial femoral condyle articulates with both the medial meniscus and medial tibial condyle, and the lateral femoral condyle articulates with both the lateral meniscus and lateral tibial condyle.

1. The prominent bony bumps that can be palpated on the medial and lateral sides of the distal end of the femur are the __medial and lateral femoral epicondyles__.

2. The prominent bony bumps that can be palpated on the medial and lateral sides of the proximal end of the tibia are the __medial and lateral tibial condyles__.

3. The prominent bony bump that can be palpated on the anterior surface of the tibia about an inch below the tibial condyles is the __tibial tuberosity__.

4. The prominent bony bump that can be palpated at the upper end of the fibula is the __head of the fibula__.

5. Which three knee injuries commonly occur when the knee is hit by a very powerful valgus force?

It is relatively common in contact sports such as football, rugby, and lacrosse for a player to be hit on the lateral aspect of the knee by the body of a second player. The momentum of the second player's body imposes a powerful, medially-directed force across the first player's knee. A medially-directed force at the knee is called a valgus force. If the first player's foot is firmly planted on the ground below, the valgus force acts to widen the medial joint space of the knee joint; the widening stretches the medial collateral ligament and the attached medial meniscus. If the valgus force is severe enough, the medial collateral ligament and medial meniscus are stretched to the extent that they abruptly tear. The complete tearing of the medial collateral ligament frees the valgus force to widen the medial joint space even more. If the medial joint space markedly widens, the anterior cruciate ligament becomes so severely stretched over the medial aspect of the lateral femoral condyle that the midportion of the ligament suffers an abrupt rupture. The net result of the severe valgus force at the knee is thus a triad of injuries: a completely torn medial collateral ligament, a ruptured anterior cruciate ligament, and a torn medical meniscus. Orthopedic surgeons call this triad of injuries the Unhappy Triad of O'Donoghue, in honor of the physician who described the basis of the three injuries.

6. Describe the locations of the 4 bursae around the anterior aspect of the knee. Which of the bursae communicate with the synovial cavity of the knee joint?

The suprapatellar bursa extends (in an adult) for about 5 to 7 cm above the upper border of the patella; it lies between the quadriceps femoris tendon and the distal end of the femur. The suprapatellar bursa is a direct extension of the knee joint's synovial cavity.

The prepatellar bursa is a subcutaneous bursa which lies between the skin and (a) the lower half of the patella and (b) the upper half of the patellar ligament. It does not communicate with the knee joint's synovial cavity.

The superficial infrapatellar bursa is a subcutaneous bursa that lies between the skin and the patellar ligament. It does not communicate with the knee joint's synovial cavity.

The deep infrapatellar bursa is an intracapsular bursa which lies between the patellar ligament and the infrapatellar fat pad (the infrapatellar fat pad lies, for the most part, between the patellar ligament and the synovial membrane of the knee joint). It does not communicate with the knee joint's synovial cavity.

7. What is the bulge sign?

The bulge sign is the most sensitive physical test for the presence of a small amount of excess fluid in the synovial cavity of the knee joint. When a small amount of excess fluid accumulates in the synovial cavity of the knee joint, the only visual evidence commonly consists of swollen soft tissues on the medial and/or lateral sides of the patellar ligament. To distinguish these swellings from swellings of the prepatellar, superficial infrapatellar, and deep infrapatellar bursae, the examiner massages the swollen soft tissues on the medial and/or lateral sides of the patellar ligament as the patient lies supine on an examination table. If most or all of the swellings on the sides of the patellar ligament represent swellings of soft tissues or bursae exterior to the knee joint's synovial cavity, then the massage will not significantly diminish the swellings. However, if most or all of the swellings represent a knee joint effusion (that is, a swelling of the knee joint's synovial cavity), then the massage will displace the excess fluid superiorly into the suprapatellar bursa and the swellings will be markedly diminished upon withdrawl of the examiner's hand. But within about a minute the soft tissues alongside the patellar ligament will swell, or bulge, again (this is the bulge sign) as the excess fluid redistributes itself back into the anteroinferior parts of the knee joint's synovial cavity alongside the patellar ligament.

8. Describe the locations of the 3 bursae around the posterior aspect of the knee. Which of the bursae communicate with the synovial cavity of the knee joint?

There are three bursae called the gastrocnemius, popliteus, and semimembranosus bursae that lie posterior to the knee joint's fibrous capsule. Each bursa lies deep to the leg or thigh muscle for which it is named. Each bursa may (but not necessarily) communicate with the knee joint's synovial cavity.

9. What is a Baker's cyst?

The expression Baker's cyst refers to a fluid-filled herniation of the synovial membrane lining the posterior aspect of the knee joint. Most Baker's cysts represent a swollen gastrocnemius or semimembranosus bursa which communicates directly with the knee joint's synovial cavity. A Baker's cyst is almost always a complication of chronic swelling of the knee joint's synovial cavity, such as may occur with rheumatoid arthritis of the knee.

10. Describe the lower limb tissues whose lymph is filtered by the lymph nodes in the back of the knee.

The lymph nodes in the back of the knee are called the popliteal lymph nodes because the region behind the knee joint is called the popliteal fossa. Popliteal lymph nodes are difficult to palpate because they lie relatively deep in the popliteal fossa; palpation of popliteal lymph nodes should be performed with the patient's leg flexed about 90° at the knee.

The popliteal lymph nodes filter lymph drained from (a) the superficial tissues on the lateral side of the foot and the posterolateral aspect of the leg and (b) the deep tissues of the foot and leg. In the limbs, the superficial tissues consist of the skin and the underlying superficial fascia. Deep tissues consist of the deep fascia and all tissues deep to the deep fascia.

Muscles of the Thigh

The muscles of the thigh all lie deep to the deep fascia of the thigh, which is called the fascia lata (Fig. 7-1). The fascia lata gives rise to extensions called intermuscular septa that attach deeply to the shaft of the femur. The fascia lata and its intermuscular septa divide the thigh muscles into three compartments called the anterior, medial, and posterior compartments.

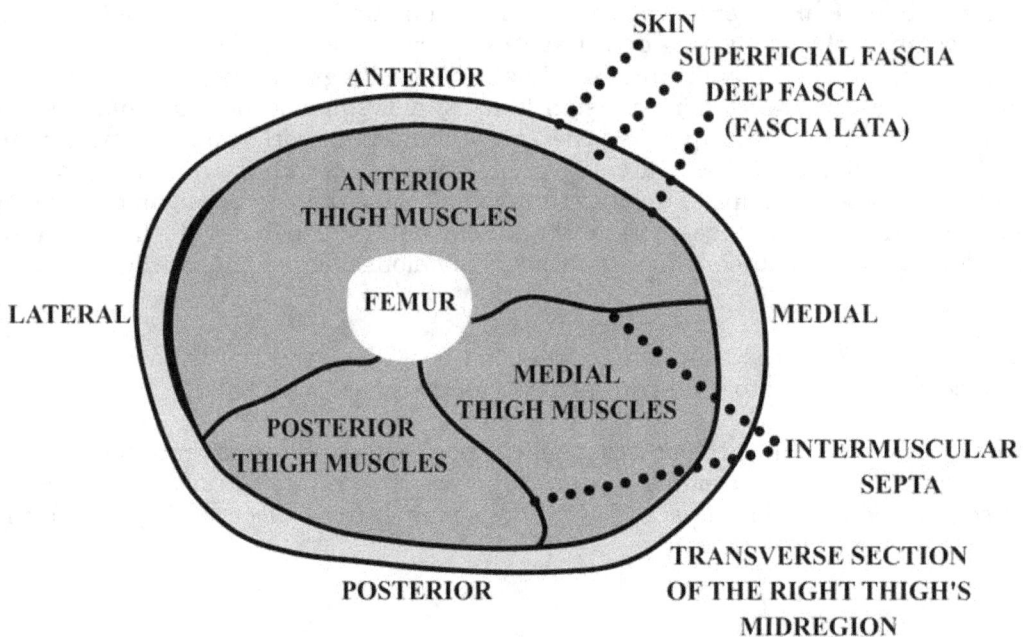

Fig. 7-1: Transverse cross section of right thigh's midregion.

All the posterior thigh muscles are innervated by either the tibial division or the common fibular division of the sciatic nerve. In the lower part of the back of the thigh, the sciatic nerve divides into its two end branches: the tibial and common fibular nerves. The tibial and common fibular divisions of the sciatic nerve are terms which refer to the nerve fibers within, respectively, the tibial and common fibular parts of the sciatic nerve.

All the medial thigh muscles, except one, are exclusively innervated by the obturator nerve.

All the anterior thigh muscles, except two, are exclusively innervated by the femoral nerve.

The 16 muscles of the thigh are listed here from A to R. Adductor magnus has two parts; each part is listed separately. Biceps femoris has two heads of origin; each head is listed separately. Refer to this list in answering questions 11-19.

A. Adductor brevis
B. Adductor longus
C. Adductor magnus
   (the adductor part)
D. Adductor magnus
   (the hamstring part)
E. Biceps femoris (the long head)
F. Biceps femoris (the short head)

G. Gracilis
H. Iliacus
I. Obturator externus
J. Pectineus
K. Psoas major
L. Rectus femoris

M. Sartorius
N. Semimembranosus
O. Semitendinosus
P. Vastus intermedius
Q. Vastus lateralis
R. Vastus medialis

11. __E__, __N__, and __O__ are the only thigh muscles which can both extend the thigh at the hip joint and flex the leg at the knee joint; these three muscles are all posterior thigh muscles and are commonly referred to as the hamstrings.

12. __L__, __P__, __Q__, and __R__ are the only thigh muscles which can extend the leg at the knee joint. These four muscles are all anterior thigh muscles and are commonly referred to as the quadriceps femoris. They all share a common tendon of insertion in which the patella is embedded.

13. __H__ and __K__ are the two anterior thigh muscles that act together as the most powerful flexors of the thigh at the hip joint. They are commonly referred to as iliopsoas.

14. __A__, __B__, __C__, __G__, and __J__ are the 5 thigh muscles which can adduct the thigh at the hip joint.

15. __E__ and __F__ are the only thigh muscles which can externally rotate the leg at the knee joint.

16. __G__, __M__, __N__, and __O__ are the only thigh muscles which can internally rotate the leg at the knee joint.

17. __D__ is the only medial thigh muscle which can extend the thigh at the hip joint.

18. __J__ can both flex and adduct the thigh at the hip joint.

19. __I__ can externally rotate the thigh at the hip joint.

The nerves and nerve fibers which innervate the thigh muscles are listed here from A to E. Refer to this list in matching each thigh muscle with its innervation.

A. Femoral nerve
B. Obturator nerve
C. Common fibular portion of the sciatic nerve
D. Tibial portion of the sciatic nerve
E. Nerve fibers from the anterior rami of L1, L2, and L3

20. __B__ Adductor brevis

21. __B__ Adductor longus

22. __B__ Adductor part of adductor magnus

23. __D__ Hamstring part of adductor magnus

24. __D__ Long head of biceps femoris

25. __C__ Short head of biceps femoris

26. __B__ Gracilis

27. __A__ Iliacus

28. __B__ Obturator externus

29. __A__ and __B__ Pectineus

30. __E__ Psoas major

31. __A__ Rectus femoris

32. __A__ Sartorius

33. __D__ Semimembranosus

34. __D__ Semitendinosus

35. __A__ Vastus intermedius

36. __A__ Vastus lateralis

37. __A__ Vastus medialis

38. In the abdomen on each side, nerve fibers from the anterior rami of several spinal nerves form a plexus, or network, of nerve fibers called the lumbar plexus. Which spinal nerves contribute nerve fibers to the lumbar plexus?

__x__ L1   _____ S1
__x__ L2   _____ S2
__x__ L3   _____ S3
__x__ L4   _____ S4
_____ L5   _____ S5

39. The lumbar plexus gives rise to the femoral nerve. Which spinal nerves contribute nerve fibers to the femoral nerve?

_____ L1   _____ S1
__x__ L2   _____ S2
__x__ L3   _____ S3
__x__ L4   _____ S4
_____ L5   _____ S5

40. The lumbar plexus gives rise to the obturator nerve. Which spinal nerves contribute nerve fibers to the obturator nerve?

_____ L1   _____ S1
__x__ L2   _____ S2
__x__ L3   _____ S3
__x__ L4   _____ S4
_____ L5   _____ S5

41. Explain why the quadriceps femoris tendon reflex test assesses spinal cord reflexes at the L2, L3, and L4 spinal cord levels.

In the quadriceps femoris tendon reflex test, a reflex hammer is used to impart a sudden stretching of the patellar ligament, which is the part of the quadriceps femoris insertion tendon that extends from the patella to the tibial tuberosity. Under normal conditions, the sudden stretching of the patellar ligament elicits reflexive contraction in all 4 muscles of the quadriceps femoris. Each muscle of the quadriceps femoris group is innervated by the femoral nerve and receives nerve fibers from the anterior rami of only L2, L3, and L4.

42. Explain why it can be said that L1 and L2 control flexion of the thigh at the hip joint.

Flexion of the thigh at the hip joint is controlled mainly by L1 and L2 nerve fibers because psoas major and iliacus, which are the two most powerful flexors of the thigh at the hip joint, each receive most of their innervation from L1 and L2 nerve fibers.

43. Explain why it can be said that L3 and L4 control extension of the leg at the knee joint.

Extension of the leg at the knee joint is controlled mainly by L3 and L4 nerve fibers because the muscles of the quadriceps femoris, which are the only muscles which can extend the leg at the knee, each receive most of their innervation from L3 and L4 nerve fibers.

44. The hamstrings (semimembranosus, semitendinosus, and the long head of biceps femoris) exert major roles in the walking gait. Explain the roles exerted by the hamstrings during (a) the terminal swing phase (TSW) of the swing period and (b) the initial contact (IC) and loading response (LR) phases of the stance period.

Semimembranosus and semitendinosus both originate from the ischial tuberosity of the coxal bone and insert onto the uppermost medial side of the tibia. The long head of biceps femoris also originates from the ischial tuberosity of the coxal bone and inserts onto the head of the fibula.

The major actions exerted by the hamstrings in the walking gait are identical to those exerted by gluteus maximus. During the TSW phase, the hamstrings assist gluteus maximus in restraining and slowing down forward movement of the thigh by pulling back on the uppermost part of the leg from the bony pelvis. During the IC and LR phases, the hamstrings assist gluteus maximus in resisting the tendency of the body to fall forward at the hip joint by pulling on the back of the bony pelvis from the uppermost part of the leg.

45. The hamstring part of adductor magnus exerts major roles in the walking gait. Explain the roles exerted by the hamstring part of adductor magnus during (a) the terminal swing phase (TSW) of the swing period and (b) the initial contact (IC) and loading response (LR) phases of the stance period.

Like the hamstrings, the hamstring part of adductor magnus also originates from the ischial tuberosity of the coxal bone. It insets onto the femur just above the medial epicondyle. The major roles exerted by the hamstring part of adductor magnus in the walking gait are identical to those exerted by the hamstrings.

46. The quadriceps femoris muscles exert major roles in the walking gait. Explain the roles exerted by the quadriceps femoris muscles during (a) the terminal swing phase (TSW) of the swing period and (b) the loading response (LR) phases of the stance period.

During the TSW phase, the quadriceps femoris muscles act to extend the leg at the knee in preparation for heel strike. During the LR phase, the quadriceps femoris muscles oppose the tendency of the leg to be flexed at the knee. The leg has a tendency to be flexed at the knee during the LR phase because upper body weight is being loaded upon the limb along a line that projects posterior to the knee joint. The projection of the upper body weight line posterior to the knee joint generates a significant flexion torque at the joint.

47. Adductor longus and gracilis act in unison to exert a major role during the early part of the initial swing phase (ISW) of the swing period. Explain this action.

In the anatomical position, neither adductor longus nor gracilis can flex the thigh at the hip joint because each muscle's insertion site lies essentially in the same coronal plane as its origin from the coxal bone. However, during the early part of the initial swing phase of the walking gait, both adductor longus and gracilis can act as flexors of the thigh because their insertion sites lie posterior to their origins from the coxal bone. Their flexor activity at the hip joint during the early part of the initial swing phase accelerates forward movement of the thigh.

The Anatomical Relationships Among the
Femoral Nerve, Femoral Artery, Femoral Vein, and Femoral Canal
in the Femoral Triangle

The inguinal ligament is the structure that serves as the boundary between the abdomen and thigh in the front of the body. The inguinal ligament is part of the tendon of insertion of a muscle in the anterior wall of the abdomen; specifically, it is the lower border of the tendon of insertion of the external oblique muscle. The lower border of external oblique's tendon of insertion is called a ligament because its ends are attached to bone, specifically, parts of the coxal bone. The inguinal ligament is attached laterally to the anterior superior iliac spine and medially to the pubic tubercle of the coxal bone.

The inguinal ligament is the upper border of a triangular region called the femoral triangle in the upper anteromedial region of the thigh. The femoral triangle is bordered medially by the medial border of adductor longus and laterally by the medial border of sartorius. Four thigh muscles form a muscular floor for the femoral triangle. Proceeding from medial to lateral, the 4 muscles are adductor longus, pectineus, psoas major, and iliacus.

Knowledge of the femoral triangle is clinically important because of the following four anatomical relationships: (1) The femoral nerve enters the thigh by passing deep to the femoral triangle's upper border, the inguinal ligament. (2) The femoral artery enters the thigh by passing deep to the inguinal ligament. (3) The femoral vein exits the thigh by passing deep to the inguinal ligament. (4) There is a very narrow passageway called the femoral canal that extends into the thigh upon passing deep to the inguinal ligament. The contents of femoral hernias enter the thigh by passing through the femoral canal. A question regarding femoral hernias and the anterior abdominal wall is presented on page 158.

The femoral artery is the chief source of blood supply to the lower limb. Knowledge of where the femoral artery can be accessed in the femoral triangle is important because the femoral artery is commonly accessed when inserting stents in narrowed coronary arteries or withdrawing a small volume of arterial blood under emergency conditions for measuring arterial blood gases.

The femoral vein is the largest vein of the lower limb; it conveys most of the blood drained from the lower limb. Knowledge of where the femoral vein can be accessed in the femoral triangle is important because the femoral vein is frequently accessed if a patient requires emergency administration of medications or fluid into the blood circulation.

The femoral artery and vein are accessed in the femoral triangle immediately below the inguinal ligament because, immediately below the inguinal ligament, the femoral nerve, femoral artery, femoral vein, and femoral canal have highly predictable relationships to each other and do not overlap each other.

48. Proceeding from lateral to medial, what are the relationships among the femoral nerve (N), femoral artery (A), femoral vein (V), and femoral canal (C) immediately below the inguinal ligament?

__x__ NAVC	_____ ANVC	_____ VNAC	_____ CNAV
_____ NACV	_____ ANCV	_____ VNCA	_____ CNVA
_____ NVAC	_____ AVNC	_____ VANC	_____ CANV
_____ NVCA	_____ AVCN	_____ VACN	_____ CAVN
_____ NCAV	_____ ACNV	_____ VCAN	_____ CVAN
_____ NCVA	_____ ACVN	_____ VCNA	_____ CVNA

The easiest way to remember the relationships among the femoral nerve (N), femoral artery (A), femoral vein (V), and femoral canal (C) immediately below the inguinal ligament is to recall the acronym NAVAL. The femoral canal contains 1 or 2 lymph nodes called deep inguinal lymph nodes and accompanying lymphatic vessels. The letters of the acronym NAVAL represent the following lateral-to-medial order of structures: femoral **N**erve, femoral **A**rtery, femoral **V**ein, **A**nd **L**ymphatics in the femoral canal.

49. Where can the pulsations of the femoral artery be palpated immediately below the inguinal ligament?

The femoral pulse is palpable immediately below the inguinal ligament at the point midway between the anterior superior iliac spine of the coxal bone and the pubic symphysis (the secondary cartilaginous joint joining the pubic bodies in the bony pelvis). In an adult, the femoral vein can be accessed about 1 cm medial to the femoral pulse.

The end segment of the great saphenous vein and two groups of superficial lymph nodes lie within the superficial fascia overlying the femoral triangle. The great saphenous vein, which is the longest vein in the human body and the largest superficial vein of the lower limb, ends just below the inguinal ligament by passing through a hole in the deep fascia of the thigh and joining the femoral vein. One of the superficial lymph node groups overlying the femoral triangle lies alongside the end segment of the great saphenous vein, this group is called the vertical group of superficial inguinal lymph nodes. The other superficial lymph node group overlying the femoral triangle is strung out in an adult about an inch below the inguinal ligament; this group is called the horizontal group of superficial inguinal lymph nodes.

50. Describe the lower limb tissues whose lymph is filtered by the vertical group of superficial inguinal lymph nodes.

The vertical group of superficial inguinal lymph nodes filters lymph drained from the same tissues whose venous blood is drained by the great saphenous vein. These tissues include all the superficial tissues of the lower limb except those of the lateral half of the foot, the posterolateral side of the leg, and the gluteal region.

51. Describe the lower limb tissues whose lymph is filtered by the horizontal group of superficial inguinal lymph nodes.

The horizontal group of superficial inguinal lymph nodes filters lymph drained from the urethra, the lower half of the anal canal, and the external genitalia of both sexes (which includes the vagina below the hymen in the female but not the testes in the male). The group also drains lymph from the anterior abdominal wall up to the level of the umbilicus and from the superficial tissues of the gluteal region.

LEG, ANKLE, AND FOOT – Part A: Questions

Dissection of the leg, ankle, and foot in gross lab focuses on identification of the 13 leg muscles, the three major nerves of the leg (the deep fibular, superficial fibular, and tibial nerves), the courses of the anterior and posterior tibial arteries through the leg and across the ankle, and the courses of the great and lesser saphenous veins through the leg. Six topics dominate the leg, ankle, and foot anatomy most frequently applied in clinical practice: (1) the sites around the ankle at which the dorsalis pedis and posterior tibial artery pulses can be palpated, (2) the spinal cord levels at which spinal cord reflexes are assessed by the Achilles tendon reflex, (3) the ankle ligament which is most frequently torn when the plantarflexed foot is forcibly supinated, (4) the innervation and major actions of the leg muscles, (5) the nature of a high steppage gait, and (6) the course by which the great saphenous vein ascends from the dorsum of the foot to the anteromedial aspect of the leg.

Bones and Joints Associated with the Actions of Leg Muscles

1. The drawing below (Fig. 8-1) shows a superior view of the outlines of the bones of the foot, which consists of the tarsals (the 7 most posterior bones of the foot), the metatarsals, and the phalanges of the toes. Print the letter of each bone and bone part listed below over the outline in the drawing. Label each joint listed.

A. Calcaneus
B. Cuboid
C. Intermediate cuneiform
D. Lateral cuneiform
E. Medial cuneiform
F. Navicular
G. Talus

H. Base of 5th metatarsal
I. Head of 1st metatarsal
J. Shaft of proximal phalanx of big toe
K. Proximal interphalangeal joint of the little toe
L. Tarsometatarsal joint of big toe
M. Calcaneonavicular joint

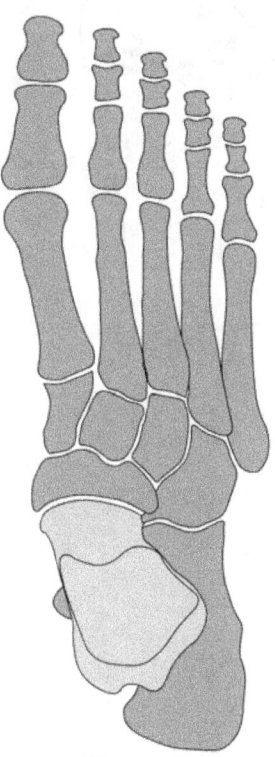

Fig. 8-1

2. The prominent bony bump that can be palpated on the medial side of the ankle is the _____.

3. The prominent bony bump that can be palpated on the lateral side of the ankle is the _____.

4. Which tarsal bone articulates with the distal ends of the tibia and fibula in the ankle joint? _____

5. The ligament which supports the medial side of the ankle joint is called the _____.

6. The three ligaments which support the lateral side of the ankle joint are called the _____.

7. Of the ligaments which support the medial and lateral sides of the ankle joint, which one is most frequently torn when the plantarflexed foot is forcibly supinated?

8. The two shaded areas in Fig. 8.5 are areas in the sole of the right foot that come into contact with the ground below when a person is standing in the anatomical position. Identify the bones or parts of bones in the foot bearing weight in these areas.

        Area A: _____

        Area B: _____

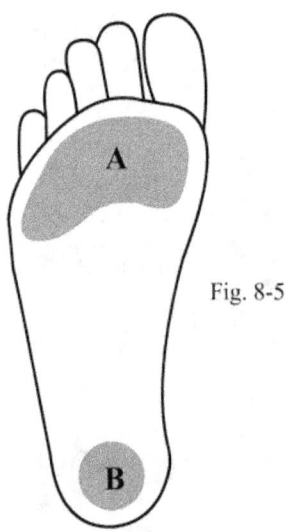

Fig. 8-5

Muscles of the Leg

The 13 muscles of the leg are listed here from A to L. Refer to this list in answering questions 9-19.

A. Extensor digitorum longus
B. Extensor hallucis longus
C. Fibularis brevis
D. Fibularis longus
E. Fibularis tertius

F. Flexor digitorum longus
G. Flexor hallucis longus
H. Gastrocnemius
I. Plantaris
J. Popliteus

K. Soleus
L. Tibialis anterior
M. Tibialis posterior

9. _____, _____, and _____ are the chief dorsiflexors of the foot.

10. _____ and _____ are the chief plantarflexors of the foot.

11. _____ and _____ are the chief supinators of the foot.

12. _____ and _____ are the chief pronators of the foot.

13. _____ is the only lower limb muscle which can extend the phalanges of the big toe.

14. _____ is the only lower limb muscle which can flex the phalanges of the big toe.

15. _____ is the only lower limb muscle which can extend the phalanges of the four lateral toes.

16. _____ is the only lower limb muscle which can flex the phalanges of the four lateral toes.

17. _____ is the muscle which is said to 'unlock' the knee joint when the leg is flexed from the fully extended position.

18. The Achilles tendon is the common tendon of insertion for _____, _____, and _____.

19. _____ is a relatively weak dorsiflexor and pronator of the foot.

The nerves which innervate the leg muscles are listed here from A to C. Refer to this list in matching each leg muscle with its innervation.

A. Deep fibular nerve
B. Superficial fibular nerve
C. Tibial nerve

20. _____ Extensor digitorum longus

21. _____ Extensor hallucis longus

22. _____ Fibularis brevis

23. _____ Fibularis longus

24. _____ Fibularis tertius

25. _____ Flexor digitorum longus

27. _____ Gastrocnemius

28. _____ Plantaris

29. _____ Popliteus

30. _____ Soleus

31. _____ Tibialis anterior

32. _____ Tibialis posterior

26. _____ Flexor hallucis longus

33. Explain why it can be said that L5 controls extension of the big toe's phalanges.

34. Explain why the Achilles tendon reflex test assesses spinal cord reflexes at the S1 and S2 spinal cord levels.

35. Explain why it can be said that S1 and S2 control plantarflexion of the foot?

36. The chief dorsiflexors of the foot (tibialis anterior, extensor hallucis longus, and extensor digitorum longus) exert major roles in the walking gait. Explain how these three muscles help define the heel rocker action during the loading response (LR) phase of the stance period and then help clear the toes from the ground at the beginning of the initial swing (ISW) phase of the swing period.

37. Describe and explain the abnormal gait that a person adopts following paralysis of the chief dorsiflexors of the foot.

38. Explain why a severe injury to the uppermost lateral side of the leg may result in foot drop.

39. The chief plantarflexors of the foot (gastrocnemius and soleus) exert major roles in the walking gait. Explain how these two muscles help define the ankle rocker action during the mid stance (MST) phase of the stance period and the forefoot rocker action during the terminal stance (TST) phase of the stance period.

Arteries and Veins of the Leg

In the lower part of the popliteal fossa, the popliteal artery divides into its two terminal branches: the anterior and posterior tibial arteries. The anterior tibial artery ends in front of the ankle joint as the origin of the dorsalis pedis (Latin for the "artery on the dorsum of the foot").

40. Where are the pulsations of dorsalis pedis palpable?

41. Where are the pulsations of the posterior tibial artery palpable?

42. Explain why the great saphenous vein is frequently used to construct a vascular bypass around a site of coronary arterial occlusion.

Sciatica

Sciatica is the term given to the syndrome in which pain occurs anywhere along the course of the sciatic nerve and its branches. The most common cause of sciatica is herniation of an intervertebral disc in the lumbar region of the spine. When an intervertebral disc in the spine's lumbar region herniates, nucleus pulposus material is typically extruded through a posterolateral sector of the anulus fibrosis. If the extruded disc material compresses the roots of one of the five spinal nerves (L4, L5, S1, S2, and S3) that contribute nerve fibers to the sciatic nerve, the person will experience sciatica, that is, pain radiating down the sciatic nerve and possibly one or more of its branches.

43. Identify the two most commonly herniated intervertebral discs in the lumbar region of the spine.

   _____ 1st lumbar disc
   _____ 2nd lumbar disc
   _____ 3rd lumbar disc
   _____ 4th lumbar disc
   _____ 5th lumbar disc

44. Identify the spinal nerve roots that are most commonly pressed upon when each of the lumbar discs identified in question 43 is herniated.

45. A physical exam procedure called the straight leg raising test is commonly conducted to explore the anatomical basis of lower back and/or lower limb pain, especially if the patient's history suggests sciatica. Describe the two parts of the straight leg raising test and explain how each part explores the anatomical basis of lower back and/or lower limb pain.

46. If both parts of the straight leg raising test yield positive findings in a patient, a second physical exam procedure called the crossed straight leg raising test is conducted to further investigate the patient's lower back and/or lower limb pain. Describe the crossed straight leg raising test and explain how the test further explores the anatomical basis of lower back and/or lower limb pain.

Meningitis

47. The brain and the spinal cord are enveloped by three membranes called meninges. The spinal meninges (the meninges which envelop the spinal cord) are direct continuations of the cranial meninges (the meninges which envelop the brain). Inflammation of the meninges is called meningitis. In most cases, inflammation is the result of infection by bacterial, viral, fungal, or parasitic agents. Bacterial meningitides (cases of bacterial meningitis) have serious sequelae (serious conditions resulting as a consequence of the disease) or a fatal outcome if not promptly treated with appropriate antibiotics.

Brudzinski's and Kernig's tests are the physical tests commonly used to elicit signs of meningitis. What signs of meningitis do Brudzinski's and Kernig's tests elicit? Describe Brudzinski's and Kernig's tests and explain how the anatomical basis by which each test elicits signs of meningitis is related to the anatomical basis by which the straight leg raising test elicits signs of sciatica.

Length of the Lower Limbs

48. As the patient lies supine upon an examination table with the lower limbs symmetrically aligned and the legs fully extended at the knees, the length of the patient's lower limbs may be compared by measuring the length of each lower limb from its medial malleolus to the _____ anterior superior iliac spine of its coxal bone
   _____ highest point of the iliac crest of its coxal bone
   _____ pubic tubercle of its coxal bone
   _____ upper margin of the pubic symphysis of the bony pelvis

END OF QUESTIONS IN PART A OF THE CHAPTER ON THE LEG, ANKLE, AND FOOT

102

Dissection of the leg, ankle, and foot in gross lab focuses on identification of the 13 leg muscles, the three major nerves of the leg (the deep fibular, superficial fibular, and tibial nerves), the courses of the anterior and posterior tibial arteries through the leg and across the ankle, and the courses of the great and lesser saphenous veins through the leg. Six topics dominate the leg, ankle, and foot anatomy most frequently applied in clinical practice: (1) the sites around the ankle at which the dorsalis pedis and posterior tibial artery pulses can be palpated, (2) the spinal cord levels at which spinal cord reflexes are assessed by the Achilles tendon reflex, (3) the ankle ligament which is most frequently torn when the plantarflexed foot is forcibly supinated, (4) the innervation and major actions of the leg muscles, (5) the nature of a high steppage gait, and (6) the course by which the great saphenous vein ascends from the dorsum of the foot to the anteromedial aspect of the leg.

Bones and Joints Associated with the Actions of Leg Muscles

1. The drawing below (Fig. 8-1) shows a superior view of the outlines of the bones of the foot, which consists of the tarsals (the 7 most posterior bones of the foot), the metatarsals, and the phalanges of the toes. Print the letter of each bone and bone part listed below over the outline in the drawing. Label each joint listed.

A. Calcaneus
B. Cuboid
C. Intermediate cuneiform
D. Lateral cuneiform
E. Medial cuneiform
F. Navicular
G. Talus

H. Base of 5th metatarsal
I. Head of 1st metatarsal
J. Shaft of proximal phalanx of big toe
K. Proximal interphalangeal joint of the little toe
L. Tarsometatarsal joint of big toe
M. Calcaneonavicular joint

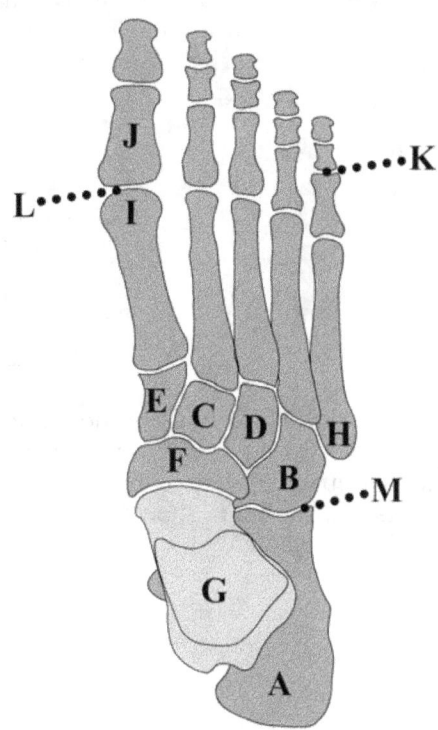

Fig. 8-1

12 of the 13 muscles of the leg move the foot as a whole; the 13th muscle rotates the tibia relative to the femur. There are 4 major movements of the foot: dorsiflexion and plantarflexion of the foot at the ankle joint and supination and pronation of the foot at the transverse tarsal and functional subtalar joints.

Fig. 8-2 shows medial views of the bones of the foot. The drawing in the middle shows a medial view of the foot in the anatomical position. Dorsiflexion of the foot is the movement in which the foot as a whole rotates in the ankle joint in such a fashion that the forefoot is raised. Plantarflexion of the foot is the movement in which the foot as a whole rotates in the ankle joint in such a fashion that the forefoot is lowered.

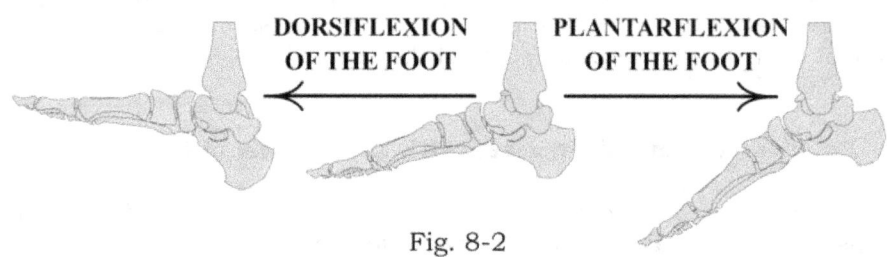

DORSIFLEXION OF THE FOOT   **PLANTARFLEXION OF THE FOOT**

Fig. 8-2

Fig. 8-3 shows that supination of the foot is basically a 50-50 combination of foot adduction (the movement by which the foot as a whole moves closer to the midline of the body) and foot inversion (the movement by which the medial edge of the foot is raised and the lateral edge of the foot is lowered). The bones of the foot are in their most tightly packed configuration when the foot is fully supinated. Fig. 8-4 shows that pronation of the foot is basically a 50-50 combination of foot abduction (the movement by which the foot as a whole moves farther from the midline of the body) and foot eversion (the movement by which the lateral edge of the foot is raised and the medial edge of the foot is lowered). The bones of the foot are in their most loosely packed configuration when the foot is fully pronated.

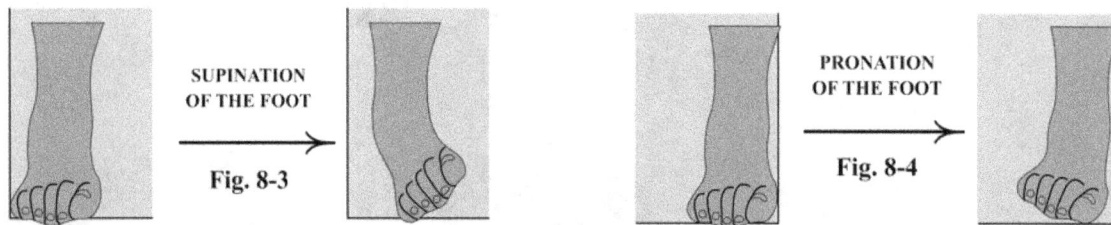

SUPINATION OF THE FOOT   Fig. 8-3   PRONATION OF THE FOOT   Fig. 8-4

The joints most responsible for the provision of supination and pronation of the foot are two functional joints called the transverse tarsal and functional subtalar joints. The transverse tarsal joint consists of (a) the articulation between the talus and navicular and (b) the articulation between the calcaneus and cuboid. The functional subtalar joint consists of the three articulations between the talus and calcaneus.

2. The prominent bony bump that can be palpated on the medial side of the ankle is the __medial malleolus__.

3. The prominent bony bump that can be palpated on the lateral side of the ankle is the __lateral malleolus__.

4. Which tarsal bone articulates with the distal ends of the tibia and fibula in the ankle joint? __talus__

5. The ligament which supports the medial side of the ankle joint is called the __deltoid ligament__.

6. The three ligaments which support the lateral side of the ankle joint are called the __anterior talofibular, calcaneofibular, and posterior talofibular ligaments__.

7. Of the ligaments which support the medial and lateral sides of the ankle joint, which one is most frequently torn when the plantarflexed foot is forcibly supinated?

   Forcible supination of a plantarflexed foot markedly stresses the lateral side of the ankle joint. The ankle ligament most frequently torn by such a stress is the anterior talofibular ligament.

8. The two shaded areas in Fig. 8.5 are areas in the sole of the right foot that come into contact with the ground below when a person is standing in the anatomical position. Identify the bones or parts of bones in the foot bearing weight in these areas.

   Area A: __heads of the metatarsals__

   Area B: __calcaneus__

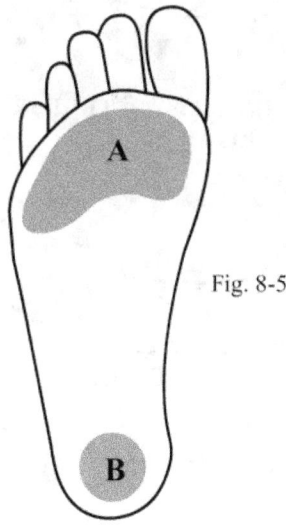

Fig. 8-5

Muscles of the Leg

The muscles of the leg all lie deep to the deep fascia of the leg, which is called the crural fascia (Fig. 8-6). The crural fascia gives rise to extensions called intermuscular septa two of which attach deeply to the shaft of the fibula. The crural fascia, its intermuscular septa, and the interossoeus membrane joining the shafts of the tibia and fibula divide the muscles of the leg into four compartments called the anterior, lateral, deep posterior, and superficial posterior compartments.

All the deep posterior and superficial posterior leg muscles are innervated by the tibial nerve.

In the uppermost part of the leg, the common fibular nerve divides into the deep and superficial fibular nerves. The deep fibular nerve innervates all the anterior leg muscles, and the superficial fibular nerve innervates all the lateral leg muscles.

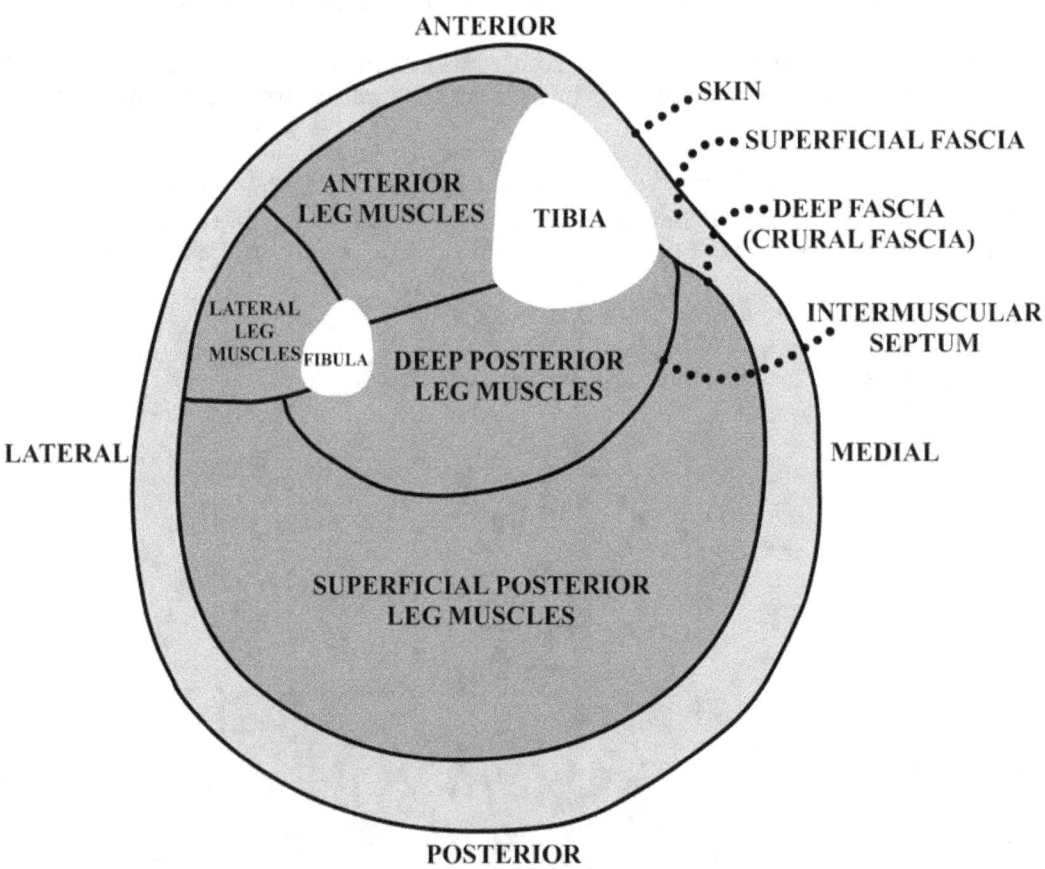

Fig. 8-6: Transverse cross section of right leg's midregion

106

The 13 muscles of the leg are listed here from A to L. Refer to this list in answering questions 9-19.

A. Extensor digitorum longus   F. Flexor digitorum longus   K. Soleus
B. Extensor hallucis longus   G. Flexor hallucis longus   L. Tibialis anterior
C. Fibularis brevis   H. Gastrocnemius   M. Tibialis posterior
D. Fibularis longus   I. Plantaris
E. Fibularis tertius   J. Popliteus

9. __A__, __B__, and __L__ are the chief dorsiflexors of the foot.

10. __H__ and __K__ are the chief plantarflexors of the foot.

11. __L__ and __M__ are the chief supinators of the foot.

12. __C__ and __D__ are the chief pronators of the foot.

13. __B__ is the only lower limb muscle which can extend the phalanges of the big toe.

14. __G__ is the only lower limb muscle which can flex the phalanges of the big toe.

15. __A__ is the only lower limb muscle which can extend the phalanges of the four lateral toes.

16. __F__ is the only lower limb muscle which can flex the phalanges of the four lateral toes.

17. __J__ is the muscle which is said to 'unlock' the knee joint when the leg is flexed from the fully extended position.

18. The Achilles tendon is the common tendon of insertion for __H__, __I__, and __K__.

19. __E__ is a relatively weak dorsiflexor and pronator of the foot.

The nerves which innervate the leg muscles are listed here from A to C. Refer to this list in matching each leg muscle with its innervation.

A. Deep fibular nerve
B. Superficial fibular nerve
C. Tibial nerve

20. __A__ Extensor digitorum longus

21. __A__ Extensor hallucis longus

22. __B__ Fibularis brevis

23. __B__ Fibularis longus

24. __A__ Fibularis tertius

25. __C__ Flexor digitorum longus

26. __C__ Flexor hallucis longus

27. __C__ Gastrocnemius

28. __C__ Plantaris

29. __C__ Popliteus

30. __C__ Soleus

31. __A__ Tibialis anterior

32. __C__ Tibialis posterior

33. Explain why it can be said that L5 controls extension of the big toe's phalanges.

Extension of the big toe's phalanges is controlled by L5 because extensor hallucis longus, which is the only lower limb muscle which can extend the phalanges of the big toe, receives all of its innervation exclusively from L5 nerve fibers.

34. Explain why the Achilles tendon reflex test assesses spinal cord reflexes at the S1 and S2 spinal cord levels.

The three muscles in the superficial posterior muscle compartment of the leg (gastrocnemius, soleus, and plantaris) share the Achilles tendon as their common tendon of insertion. Under normal conditions, the sudden stretching of the Achilles tendon elicits reflexive contraction in all three of these superficial posterior leg muscles. Each of these three muscles is innervated by the tibial nerve and receives nerve fibers from the anterior rami of only S1 and S2.

35. Explain why it can be said that S1 and S2 control plantarflexion of the foot.

Plantarflexion of the foot is controlled by S1 and S2 because gastrocnemius and soleus, which are the most powerful plantarflexors of the foot, are each innervated exclusively by S1 and S2 nerve fibers.

36. The chief dorsiflexors of the foot (tibialis anterior, extensor hallucis longus, and extensor digitorum longus) exert major roles in the walking gait. Explain how these three muscles help define the heel rocker action during the loading response (LR) phase of the stance period and then help clear the toes from the ground at the beginning of the initial swing (ISW) phase of the swing period.

During the LR phase of the stance period, the chief dorsiflexors of the foot act to resist a plantarflexion torque at the ankle joint (which is a consequence of the projection of the upper body weight line posterior to the ankle joint). This resistance to plantarflexion not only controls and slows the plantarflexion that brings the sole of the foot into full contact with the surface below but also pulls the leg forward at the ankle as the foot is slowly plantarflexed. These two effects help define the heel rocker action that occurs during the loading response phase.

During the first half of the swing period, the chief dorsiflexors of the foot act to dorsiflex the foot from its plantarflexed position at toe off to an almost neutral position by the mid swing phase; this activity helps clear the toes from the ground immediately following toe off. Their continued activity during the remainder of the swing period maintains the ankle in an almost neutral position until heel strike.

37. Describe and explain the abnormal gait that a person adopts following paralysis of the chief dorsiflexors of the foot.

Paralysis or profound weakness of the chief dorsiflexors of the foot results in a disability called foot drop. The afflicted person cannot dorsiflex the foot during the first half of the swing period nor maintain the ankle in an almost neutral position during the latter half of the swing period in preparation for heel strike. In other words, because the foot cannot be dorsiflexed during the swing period, it appears that the forefoot drops; hence, the name of the disability. The consequence of this disability is that, if the person attempts to walk normally, the toes of the affected foot strike the ground at some point during the swing period.

A person can compensate for the loss of foot dorsiflexion action by raising the foot higher than normal (through increased flexion of the thigh at the hip) during the swing period. The exaggerated raising of the foot during the swing period produces an abnormal gait called a high steppage gait. The high steppage gait permits almost the entire sole of the affected foot to strike the ground at the beginning of the initial contact phase.

38. Explain why a severe injury to the uppermost lateral side of the leg may result in foot drop.

As the common fibular nerve extends from the popliteal fossa into the leg, it winds around the lateral side of the neck of the fibula, where it lies deep to only the skin and superficial fascia and thus is relatively exposed to injury. In the vicinity immediately below the neck of the fibula, the common fibular nerve divides into the deep and superficial fibular nerves. The deep fibular nerve innervates all the chief dorsiflexors of the foot. A severe injury to the uppermost lateral side of the leg may thus result in foot drop because it puts at risk of injury the nerve fibers in the common fibular nerve that innervate the chief dorsiflexors of the foot.

39. The chief plantarflexors of the foot (gastrocnemius and soleus) exert major roles in the walking gait. Explain how these two muscles help define the ankle rocker action during the mid stance (MST) phase of the stance period and the forefoot rocker action during the terminal stance (TST) phase of the stance period.

Shortly after the beginning of the MST phase, the upper body weight line changes from an alignment in which it projects posterior to the ankle to an alignment in which it projects anterior to the ankle. This change in the alignment of the upper body weight line pulls the leg forward at the ankle joint. Gastrocnemius and soleus, however, resist this forward movement of the leg at the ankle joint by pulling backward on the back of the leg from their insertion onto the calcaneus in the foot. This backward pull on the leg at the ankle joint provided by gastrocnemius and soleus during the MST phase controls and slows the forward movement of the leg at the ankle joint. Gastrocnemius and soleus thus help define the ankle rocker action during the MST phase through their restraint of forward movement of the leg at the ankle joint.

At the end of the MST phase, the backward pull on the leg provided by gastrocnmeius and soleus restrains further forward movement of the leg at the ankle joint. At this point, this restraint in combination with the continued forward pull exerted on the leg by the advancing body leads to heel rise, the movement in which the hindfoot and midfoot roll upward and forward around the metatarsophalangeal joints. Heel rise marks the beginning of the TST phase and initiates the forefoot rocker action in which the lower limb rolls forward on the forefoot. Gastrocnemius and soleus thus help define the forefoot rocker action during the TST phase through their continued restraint of forward movement of the leg at the ankle joint. The forward movement of the upper body beyond the supporting foot and the roll of the body around the forefoot rocker during the TST phase combine to force the body to fall freely forward. This free forward fall is the principal force that propels the body forward during the walking gait.

Arteries and Veins of the Leg

In the lower part of the popliteal fossa, the popliteal artery divides into its two terminal branches: the anterior and posterior tibial arteries. The anterior tibial artery ends in front of the ankle joint as the origin of the dorsalis pedis (Latin for the "artery on the dorsum of the foot").

40. Where are the pulsations of dorsalis pedis palpable?

    The pulsations of dorsalis pedis can be palpated on the dorsum of the foot immediately lateral to the tendon of extensor hallucis longus.

41. Where are the pulsations of the posterior tibial artery palpable?

    The posterior tibial artery descends from the leg into the foot by passing behind the medial malleolus. The pulsations of the posterior tibial artery can be palpated posteroinferiorly to the medial malleolus (the pulsations are most easily palpated if the foot is both dorsiflexed and inverted).

42. Explain why the great saphenous vein is frequently used to construct a vascular bypass around a site of coronary arterial occlusion.

    The great saphenous vein begins as the medial extension of the dorsal venous arch of the foot, and curves upward into the anteromedial aspect leg by passing in front of the medial malleolus. Coronary artery bypass surgery is a surgical procedure in which segments of a patient's great saphenous vein are frequently used to construct a vascular bypass around a site of coronary arterial occlusion. The great saphenous vein is commonly used for coronary artery bypass surgery because (a) it can be readily resected from the superficial fascia of the leg, (b) its diameter closely approximates that of the coronary arteries and their major branches, and (c) like the walls of the coronary arteries, its wall bears a comparatively high content of elastic tissue.

Sciatica

Sciatica is the term given to the syndrome in which pain occurs anywhere along the course of the sciatic nerve and its branches. The most common cause of sciatica is herniation of an intervertebral disc in the lumbar region of the spine. When an intervertebral disc in the spine's lumbar region herniates, nucleus pulposus material is typically extruded through a posterolateral sector of the anulus fibrosis. If the extruded disc material compresses the roots of one of the five spinal nerves (L4, L5, S1, S2, and S3) that contribute nerve fibers to the sciatic nerve, the person will experience sciatica, that is, pain radiating down the sciatic nerve and possibly one or more of its branches.

43. Identify the two most commonly herniated intervertebral discs in the lumbar region of the spine.
_____ 1st lumbar disc
_____ 2nd lumbar disc
_____ 3rd lumbar disc
__x__ 4th lumbar disc
__x__ 5th lumbar disc

44. Identify the spinal nerve roots that are most commonly pressed upon when each of the lumbar discs identified in question 43 is herniated.

When the 4th lumbar disc, which is the disc between the 4th and 5th lumbar vertebrae, herniates, the extruded material most commonly presses upon the roots of L5. When the 5th lumbar disc, which is the disc between the 5th lumbar vertebra and the sacrum, herniates, the extruded material most commonly presses upon the roots of S1.

45. A physical exam procedure called the straight leg raising test is commonly conducted to explore the anatomical basis of lower back and/or lower limb pain, especially if the patient's history suggests sciatica. Describe the two parts of the straight leg raising test and explain how each part explores the anatomical basis of lower back and/or lower limb pain.

The straight leg raising, or SLR, test is conducted with the lower limb on the painful side of the body as the patient lies supine on an examination table. For the purpose of our discussion here, let us assume that the patient has pain in the right lower limb that extends from the buttock into the back of the leg.

In the first of the two parts of the SLR test, the examiner flexes the right thigh at the hip with the leg extended at the knee in an attempt to elicit or intensify lower limb pain. The anatomical basis of the first part of the SLR test is as follows: In the supine position, the sciatic nerve on the right side of the body and the spinal nerves from which it arises are slack as they extend from the spine through the buttock into the thigh. Flexion of the thigh at the hip from 0° to 35° flexion (with the leg fully extended at the knee and the foot in a neutral position at the ankle) stretches the sciatic nerve primarily along its course through the buttock. Flexion to 35° takes up almost all the slack in the sciatic nerve along its course in the buttock and thigh. Flexion greater than 35° places increased tension on the sciatic nerve. However, from 35° to 70° flexion, almost all of the increased tension is exacted in the spine upon the roots of the spinal nerves (L4, L5, S1, S2, and S3) that contribute to the sciatic nerve. The greatest traction is exerted upon the L5 and S1 spinal nerves along their passage through their respective intervertebral foramina and the roots of these two spinal nerves along their descent in the vertebral canal (the roots of L4, S2, and S3 are only slightly stretched). If flexion of the thigh at the hip joint elicits or intensifies lower limb pain within the 35° to 70° arc of thigh flexion, then there is roughly a 50% probability that a herniated intervertebral disc is pressing upon the dural coverings on the roots of the L5 or S1 spinal nerves. The dural coverings are implicated because receptors sensitive to stretching reside in the dural coverings but not in the spinal nerves or the roots themselves.

In the second part of the SLR test, the examiner lowers the right lower limb to the highest level at which pain is relieved and then dorsiflexes the right foot at the ankle joint. The second part of the SLR test has to be conducted because there are mechanisms other than pressure upon the roots of a spinal nerve that can intensify lower limb pain during the first part of the SLR test. Almost all of these other mechanisms require movement in joints of the lumbosacral spine, the sacroiliac joint, or the hip joint in order to intensify pain within

111

the 35° to 70° arc of thigh flexion. The second part of the SLR test eliminates consideration of these other mechanisms because it does not involve movement in joints of the lumbosacral spine, the sacroiliac joint, or the hip joint. The second part of the SLR test increases traction upon the roots of the L5 and S1 spinal nerves by stretching the tibial nerve along its passage behind and below the medial malleolus when the foot is dorsiflexed at the ankle joint. Consequently, if the second part of the SLR test elicits or intensifies lower limb pain, then there is roughly an 80% probability that a herniated intervertebral disc is pressing upon the dural coverings on the roots of the L5 or S1 spinal nerves.

46. If both parts of the straight leg raising test yield positive findings in a patient, a second physical exam procedure called the crossed straight leg raising test is conducted to further investigate the patient's lower back and/or lower limb pain. Describe the crossed straight leg raising test and explain how the test further explores the anatomical basis of lower back and/or lower limb pain.

The crossed SLR test is identical to the first part of the SLR test except that (a) the test is conducted on the painless side of the body and (b) the painless lower limb is raised in an attempt to elicit or intensify pain in the painful lower limb. In other words, if the patient presents with pain in the right lower limb, the left lower limb is raised during the crossed SLR test. If the crossed SLR test elicits or intensifies lower limb pain on the painful side within the 35° to 70° arc of thigh flexion on the painless side, then the finding is almost pathognomonic of the presence of a herniated intervertebral disc in the lumbar region of the spine. The term pathognomonic means that the diagnosis of a herniated intervertebral disc has a 99% probability of being correct.

When the thigh of the painless lower limb is flexed from 35° to 70° at the hip during the crossed SLR test, the traction on the roots of the L5 and S1 spinal nerves on the painless side of the body pulls the L5 and S1 roots on both the painless and painful sides toward the painless side. If most of the extruded material of a herniated disc lies immediately medial to the L5 or S1 roots on the painful side, the movement of these roots toward the extruded material imposes traction on the roots and thereby elicits or intensifies pain in the painful lower limb. Therefore, a positive finding with the crossed SLR test is almost pathognomonic of the presence of a herniated disc whose extruded material lies immediately medial to the roots of a spinal nerve. It is important to recognize that if the extruded material of the herniated disc lay for the most part immediately lateral to the L5 or S1 roots on the painful side, the crossed SLR test will not produce a positive finding; in other words, it will not elicit or intensify pain in the painful lower limb. This is because the movement of the L5 and S1 roots on the painful side away from the extruded material will reduce traction on the roots and thereby either relieve or not affect the pain in the painful lower limb. Consequently, a negative finding with the crossed SLR test neither increases nor decreases the probability that the patient has a herniated disc; the negative finding simply indicates that there is no evidence of a herniated disc whose extruded material lies immediately medial to the roots of a spinal nerve.

Meningitis

47. The brain and the spinal cord are enveloped by three membranes called meninges. The spinal meninges (the meninges which envelop the spinal cord) are direct continuations of the cranial meninges (the meninges which envelop the brain). Inflammation of the meninges is called meningitis. In most cases, inflammation is the result of infection by bacterial, viral, fungal, or parasitic agents. Bacterial meningitides (cases of bacterial meningitis) have serious sequelae (serious conditions resulting as a consequence of the disease) or a fatal outcome if not promptly treated with appropriate antibiotics.

Brudzinski's and Kernig's tests are the physical tests commonly used to elicit signs of meningitis. What signs of meningitis do Brudzinski's and Kernig's tests elicit? Describe Brudzinski's and Kernig's tests and explain how the anatomical basis by which each test elicits signs of meningitis is related to the anatomical basis by which the straight leg raising test elicits signs of sciatica.

Brudzinski's and Kernig's tests are each conducted as the patient lies supine on an examination table with the hands folded behind the head. In Brudzinski's test, the examiner requests the patient to raise the head up, an act which stretches the spinal meninges from the cervical part of the spine. If the meninges are inflamed, such stretching produces pain in the head, neck, or back and elicits involuntary flexion of the lower limbs at the hip and knee joints. The involuntary movements at the hip and knee joints constitute a positive sign for Brudzinski's test, since they represent an attempt by the patient to minimize tension in the spinal meninges through flexion of the thighs and legs.

The mechanism by which flexion of the thighs and legs minimizes tension in the spinal meninges is related to the course of the sciatic nerve and its terminal branches in the lower limb. The sciatic nerve extends from the buttock into the thigh by passing behind the hip joint, and its two large terminal branches, the tibial and common fibular nerves, extend from the thigh into the leg by passing behind the knee joint. Whereas flexion of the thigh at the hip joint stretches the sciatic nerve along its course behind the hip joint, flexion of the leg at the knee joint slackens the tibial and common fibular nerves along their courses behind the knee joint. Combined flexion at both joints produces a net slack in the sciatic nerve and its tibial and common fibular branches; this slack extends back to the spinal nerves (L4, L5, S1, S2, and S3) which contribute nerve fibers to the sciatic nerve. It is in this fashion that flexion of the thighs and legs slackens the spinal meninges from the lumbosacral part of the spine.

In Kernig's test, the patient slowly raises one of the lower limbs, taking care to keep the leg fully extended at the knee joint. This action stretches the spinal meninges from the lumbosacral part of the spine. If such stretching produces pain in the head, neck, or back, and the pain can be relieved by flexion of the leg at the knee joint, then the test is judged positive.

Length of the Lower Limbs

48. As the patient lies supine upon an examination table with the lower limbs symmetrically aligned and the legs fully extended at the knees, the length of the patient's lower limbs may be compared by measuring the length of each lower limb from its medial malleolus to the __x__ anterior superior iliac spine of its coxal bone
    _____ highest point of the iliac crest of its coxal bone
    _____ pubic tubercle of its coxal bone
    _____ upper margin of the pubic symphysis of the bony pelvis

CHEST (THORAX) – Part A: Questions

Dissection of the chest (thorax) in gross lab focuses on identification of
  (1) the sternum, ribs, costal cartilages, intercostal muscles, and intercostal arteries, veins, and nerves,
  (2) the heart's pericardial sac, surfaces, chambers, and valves,
  (3) the coronary arteries, cardiac veins, and coronary sinus,
  (4) the lobes of the lungs,
  (5) the trachea and the main stem and lobar bronchi of the lungs,
  (6) the branches of the ascending aorta, aortic arch, and descending thoracic aorta,
  (7) the brachiocephalic veins and the superior and inferior venae cavae,
  (8) the pulmonary trunk, ligamentum arteriosum, pulmonary arteries, and pulmonary veins,
  (9) the azygous vein and its tributaries,
  (10) the paired phrenic, paired vagus, and left recurrent laryngeal nerves, and
  (11) the esophagus, thoracic duct, and diaphragm.

The thoracic anatomy most frequently applied in clinical practice is the surface anatomy relevant to the physical examination of the heart and lungs.

Rib Cage

1. Describe the anatomical relationship in the anterior chest wall that permits quick identification of the ribs and the intercostal spaces in the anterior part of the rib cage.

2. List all the ribs whose costal cartilages contribute to the costal margins of the rib cage.
   _____ 6th
   _____ 7th
   _____ 8th
   _____ 9th
   _____ 10th
   _____ 11th
   _____ 12th

3. Explain why the 1st ribs are not palpable in the anterior chest wall.

4. If a person is suffering from an abnormal collection of fluid or air in the pleural space around a lung, a chest tube may inserted through an intercostal space along the midaxillary line into the pleural space to remove the fluid or air. Explain why it is important to insert the chest tube **up and over the lower rib** bordering the intercostal space.

5. The nerve supply of the chest wall is provided by 11 pairs of intercostal nerves. Describe the derivation of the intercostal nerves from thoracic spinal nerves.

6. The 1st through 9th intercostal spaces are each supplied by anterior and posterior intercostal arteries (the 10th and 11th intercostal spaces are each supplied by only a posterior intercostal artery). Describe the origins of the anterior and posterior intercostal arteries.

7. Each intercostal space is drained by anterior and posterior intercostal veins. Describe the pathways by which blood drained by the anterior and posterior intercostal veins returns back to the right atrium of the heart.

8. The prominence of the highest, readily palpable spinous process in the midline of the back can be enhanced by asking the patient to flex his/her head forward at the neck. This is the spinous process of the _____ vertebra.

_____ 4th cervical

_____ 5th cervical

_____ 6th cervical

_____ 7th cervical

_____ 1st thoracic

_____ 2nd thoracic

_____ 3rd thoracic

_____ 4th thoracic

9. The 12th rib can be palpated in the back. True or False

10. In the back, the medial end of the spine of the scapula lies at the level of the spinous process of the _____ vertebra.

_____ 7th cervical

_____ 1st thoracic

_____ 2nd thoracic

_____ 3rd thoracic

_____ 4th thoracic

_____ 5th thoracic

_____ 6th thoracic

11. In the back, the inferior angle of the scapula lies at the level of the _____ rib or its intercostal space.

_____ 5th

_____ 6th

_____ 7th

_____ 8th

_____ 9th

_____ 10th

12. What is the common anatomical feature of thoracic vertebrae that results in the tip of the spinous process of each thoracic vertebra lying approximately at the level of the body of the vertebra below?

13. As a general rule, the anterior end of a rib lies at the level of the posterior end of the _____ rib below it.

_____ 1st

_____ 2nd

_____ 3rd

_____ 4th

_____ 5th

_____ 6th

_____ 7th

Diaphragm

14. Fig. 9-1 presents 12 different surface projections of the diaphragm's domes and central tendon to the anterior chest wall. Select the drawing which most accurately represents the anterior surface projection of the diaphragm's domes and central tendon in a healthy patient lying supine on an examination table and breathing quietly.

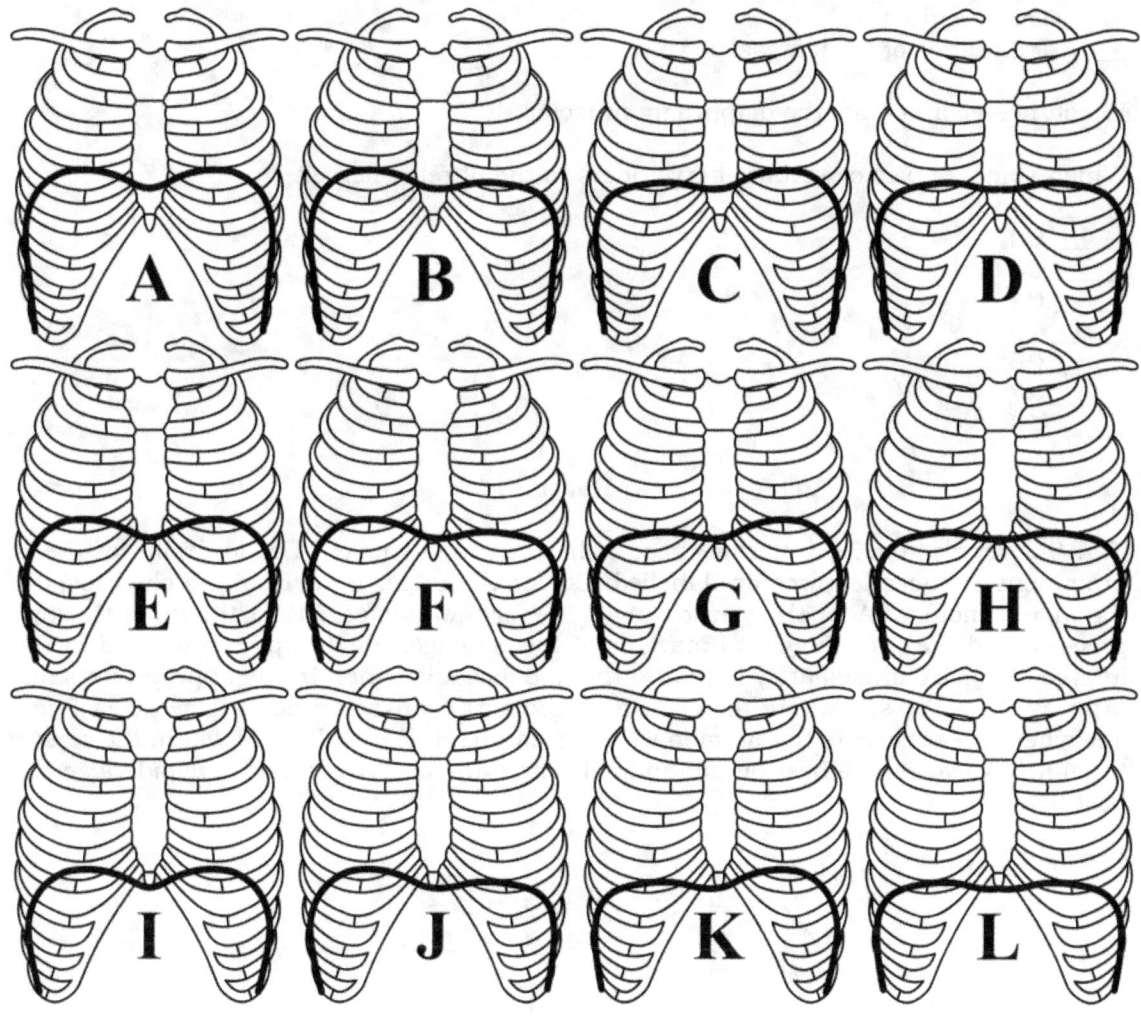

Fig. 9-1

15. There are three major openings in the diaphragm for transmission of structures between the thorax and abdomen. Each opening is named for the largest structure which passes through it. Indicate the vertebral level at which the aortic (A), caval (C), and esophageal (E) openings lie when a person is seated upright and breathing quietly.

_____ body of 6th thoracic vertebra   _____ body of 1st lumbar vertebra
_____ body of 7th thoracic vertebra   _____ body of 2nd lumbar vertebra
_____ body of 8th thoracic vertebra
_____ body of 9th thoracic vertebra
_____ body of 10th thoracic vertebra
_____ body of 11th thoracic vertebra
_____ body of 12th thoracic vertebra

16. The aortic (A), caval (C), esophageal (E) openings of the diaphragm each transmit two or more structures. Identify the opening through which each of the following structures passes:

_____ Azygos vein
_____ Branches of the left gastric artery
_____ Left and right vagus nerves
_____ Right phrenic nerve
_____ Thoracic duct
_____ Tributaries of the left gastric vein

17. Which nerves innervate the diaphragm muscle?

18. Which spinal nerves contribute nerve fibers to the phrenic nerve?

_____ C1
_____ C2
_____ C3
_____ C4
_____ C5
_____ C6
_____ C7
_____ C8

Heart

19. Fig. 9-2 represents the anterior chest wall of an adult male. The locations of the most prominent surfaces features are labelled: C's for the upper borders of the clavicles, JN for the jugular notch, SA for the sternal angle, N's for the nipples, XP for the tip of the xiphoid process, and CM's for the costal margins of the rib cage. Each nipple is located in the 4th intercostal space immediately lateral to the midclavicular line. In the appropriate squares, print an A for the site where the closure of the aortic valve can be best heard, a P for the site where the closure of the pulmonary valve can be best heard, an M where the closure of the mitral valve can be best heard, and a T where the closure of the tricuspid valve can be best heard.

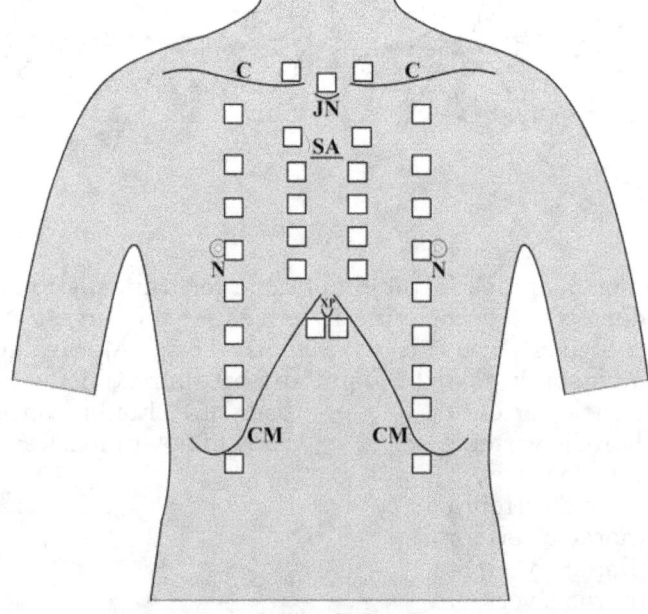

Fig. 9-2

118

20. Fig. 9-3 presents 9 different heart locations in an adult with a normal sized heart. As in Fig. 9-2 on the preceding page, each nipple is located in the 4th intercostal space immediately lateral to the midclavicular line. Select the drawing with the most accurate heart location in an adult patient with a normal sized heart who is seated upright and breathing quietly as he/she is being examined.

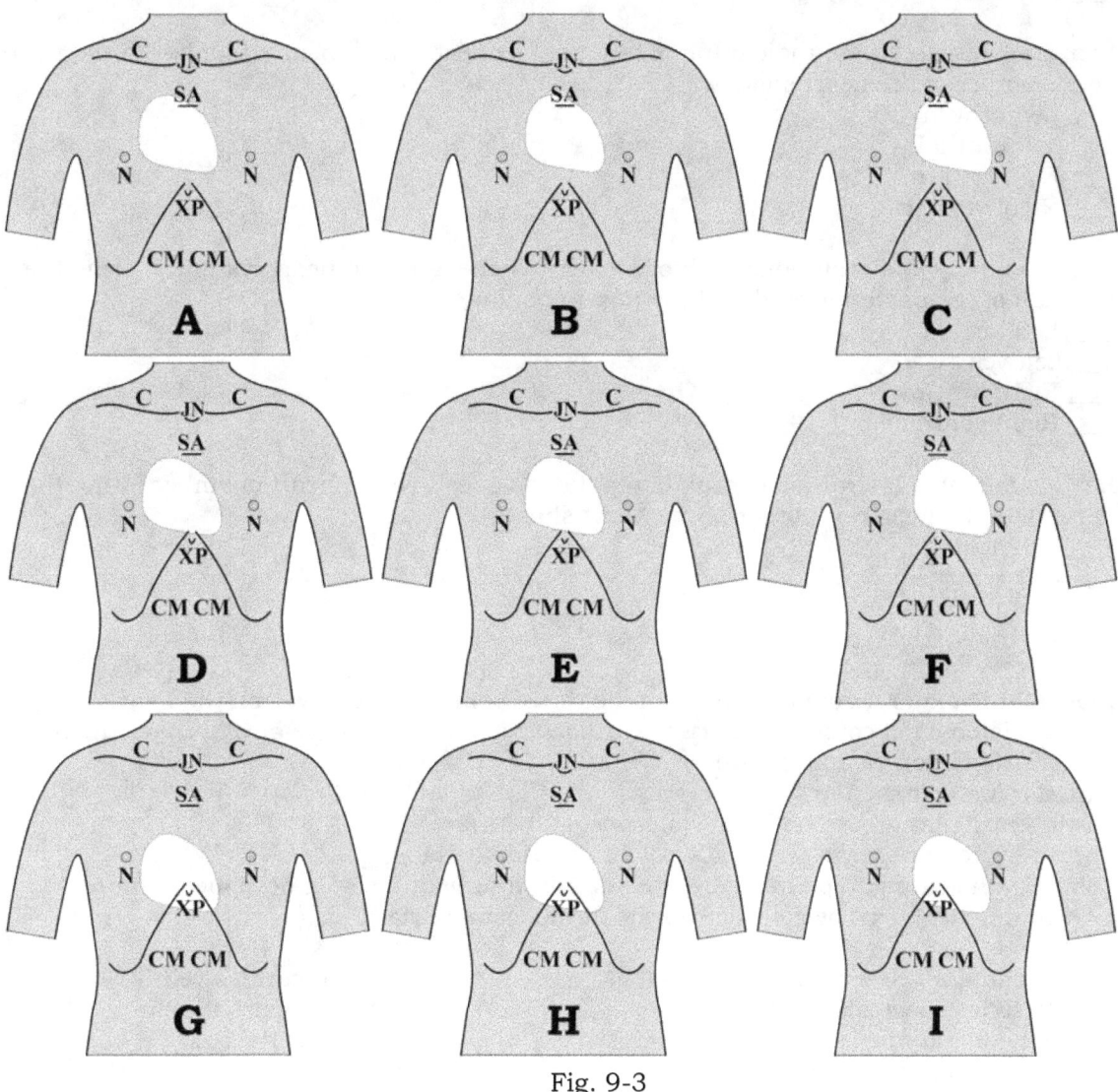

Fig. 9-3

21. Which heart chamber forms the apex of the heart?
_____ Left atrium   _____ Right atrium
_____ Left ventricle   _____ Right ventricle

22. When examining a PA chest film, which heart chamber forms most of the left border of the heart shadow?
_____ Left atrium   _____ Right atrium
_____ Left ventricle   _____ Right ventricle

119

23. When examining a PA chest film, which heart chamber forms the right border of the heart shadow?
_____ Left atrium
_____ Left ventricle
_____ Right atrium
_____ Right ventricle

24. When examining a lateral radiograph of a patient's chest, which heart chamber forms the anterior border of the heart shadow?
_____ Left atrium
_____ Left ventricle
_____ Right atrium
_____ Right ventricle

25. When examining a lateral radiograph of a patient's chest, which heart chamber forms the upper half of the posterior border of the heart shadow?
_____ Left atrium
_____ Left ventricle
_____ Right atrium
_____ Right ventricle

26. When examining a lateral radiograph of a patient's chest, which heart chamber forms the lower half of the posterior border of the heart shadow?
_____ Left atrium
_____ Left ventricle
_____ Right atrium
_____ Right ventricle

27. Almost all the major coronary arteries and cardiac veins lie in grooves on the heart's surface. Match each coronary artery or cardiac vein with the groove or path it occupies.
A: Anterior interventricular groove
B: Inferior interventricular groove
C: Atrioventricular groove on heart's sternocostal surface
D: Atrioventricular groove on heart's posteroinferior surface
E: Short path leading to point where the anterior interventricular groove meets the atrioventricular groove on the heart's sternocostal surface

_____ Left anterior descending artery (LAD)   _____ Great cardiac vein
_____ Posterior descending artery (PDA)   _____ Middle cardiac vein
_____ Left coronary artery   _____ Coronary sinus
_____ Right coronary artery

28. Which coronary artery generally supplies both the SA and AV nodes of the heart's conducting system?
_____ LAD   _____ PDA   _____ Right coronary artery

29. Fig. 9-4 represents the anterior chest wall of an adult male. The locations of the most prominent surfaces features are labelled: C's for the upper borders of the clavicles, JN for the jugular notch, SA for the sternal angle, N's for the nipples, XP for the tip of the xiphoid process, and CM's for the costal margins of the rib cage. Each nipple is located in the 4th intercostal space immediately lateral to the midclavicular line. If this drawing represented the chest of an adult male who has just suffered a heart attack, mark the box with an "X" where you would press down on the chest to apply CPR (cardiopulmonary resuscitation). Explain the anatomical basis of your decision.

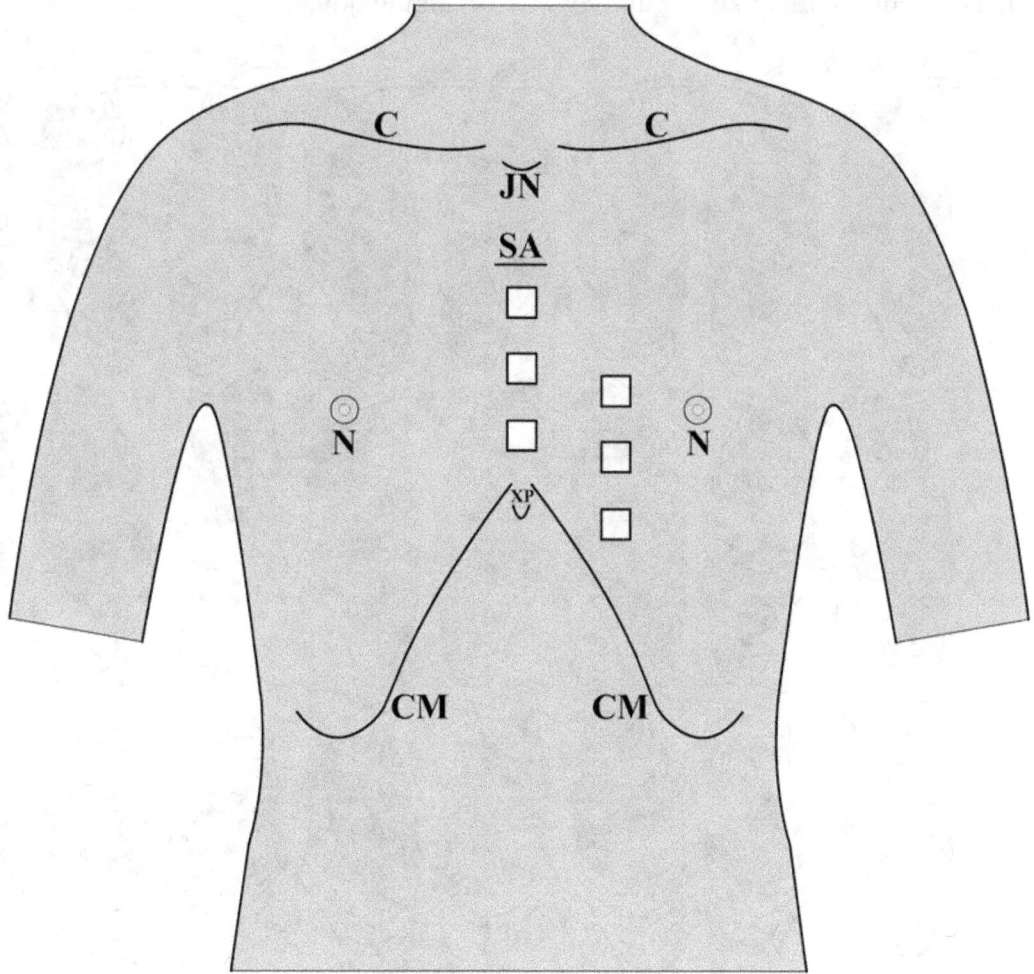

Fig. 9-4

30. Describe blood flow through the heart and the timing of the opening and closing of the heart's valves during the cardiac cycle.

Lungs and Pleura

31. Describe the parenchymal elements of the lungs and their functions.

32. Describe the mechanics of respiration. In particular, discuss the forces which expand the lungs when you breathe in and the forces which retract the lungs when you breathe out.

33. When conducting a physical examination of a patient or performing a clinical procedure anatomically related to the lungs and their pleural spaces (such as inserting a chest tube or inserting a central line into the subclavian vein), it is important to know the bony and cartilaginous structures of the rib cage which mark the surface projections of the lungs, their lobes, and pleural spaces. Fig. 9-5 presents drawings of 16 different surface projections of the lungs (not the pleural spaces) to the anterior chest wall. Select the drawing which most accurately represents the anterior surface projections of the lungs in a healthy adult patient sitting upright and breathing quiet

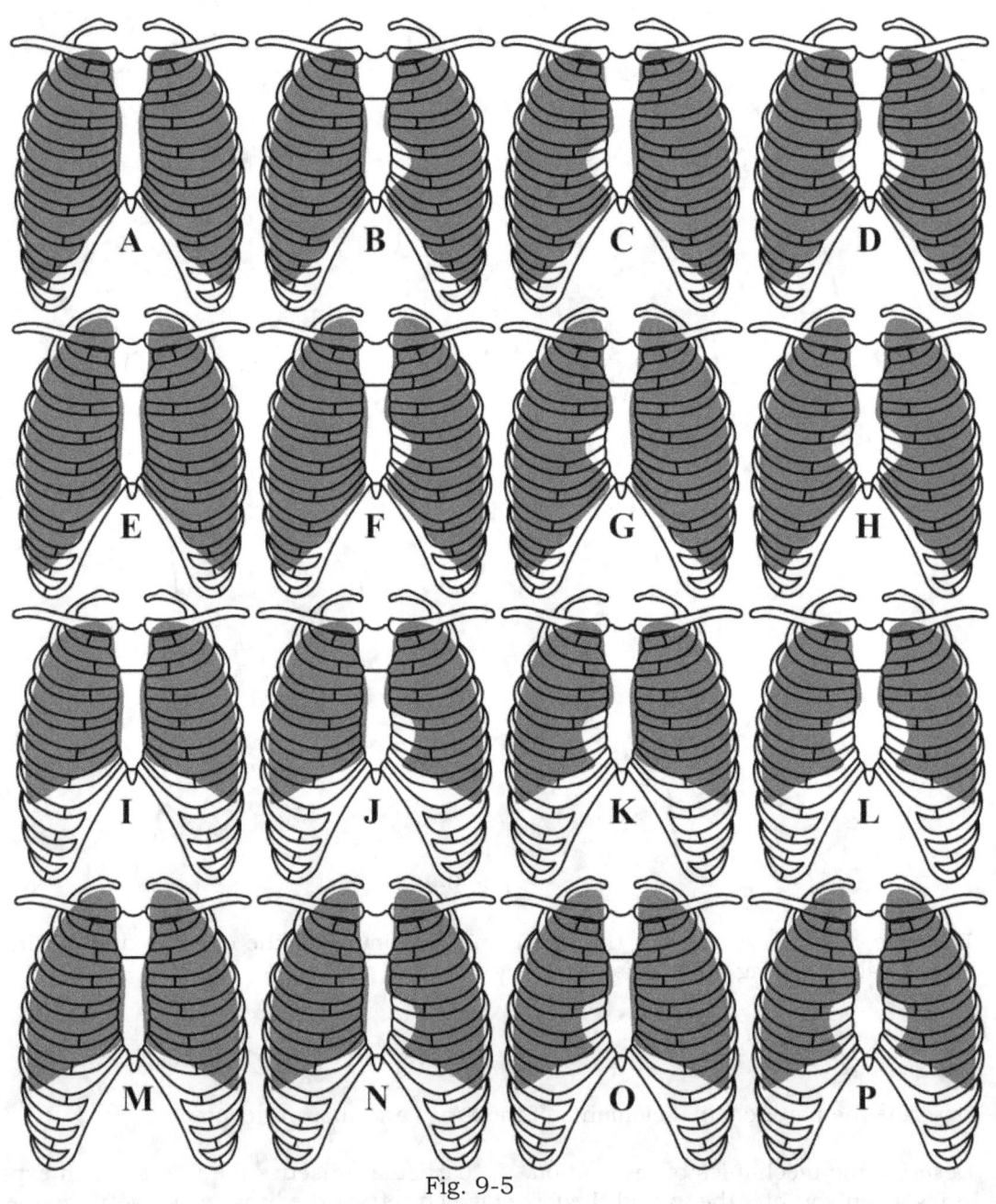

Fig. 9-5

34. Fig. 9-6 presents drawings of 3 different surface projections of the lungs (not the pleural spaces) to the posterior chest wall. Select the drawing which most accurately represents the posterior surface projections of the lungs in a healthy adult patient sitting upright and breathing quietly.

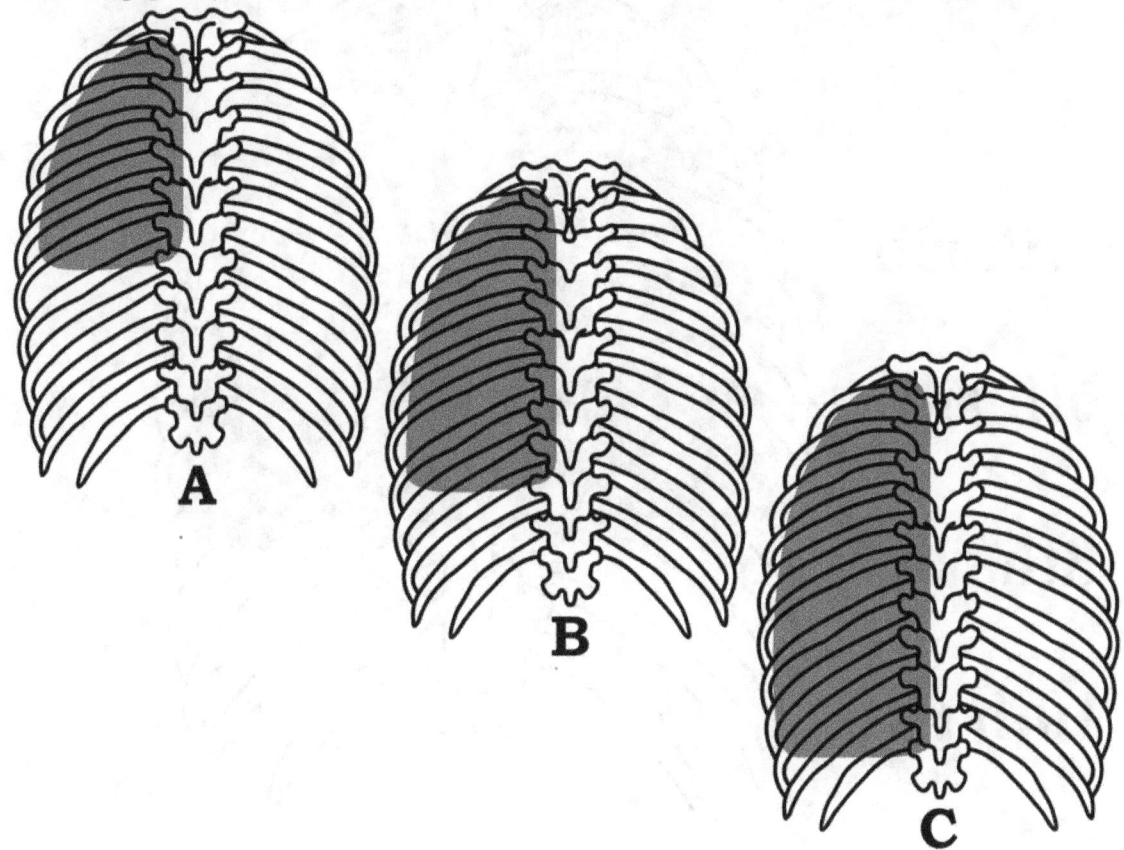

Fig. 9-6

35. Based upon your answers to questions 33 and 34, complete the following statement:
When a person is seated upright and breathing slowly,
the midclavicular line intersects the lower margin of either lung
at the level of the _____ rib,
the midaxillary line intersects the lower margin of the lung
at the level of the _____ rib, and
the lateral border of the spine intersects the lower margin of the lung
at the level of the _____ rib.

36. The oblique fissures of both lungs project to the posterior chest wall. Fig. 9-7 presents 4 different surface projections, labelled A through D, of the left lung's oblique fissure to the posterior chest wall. Select the drawing which most accurately represents the posterior surface projection of the left lung's oblique fissure to the posterior chest wall.

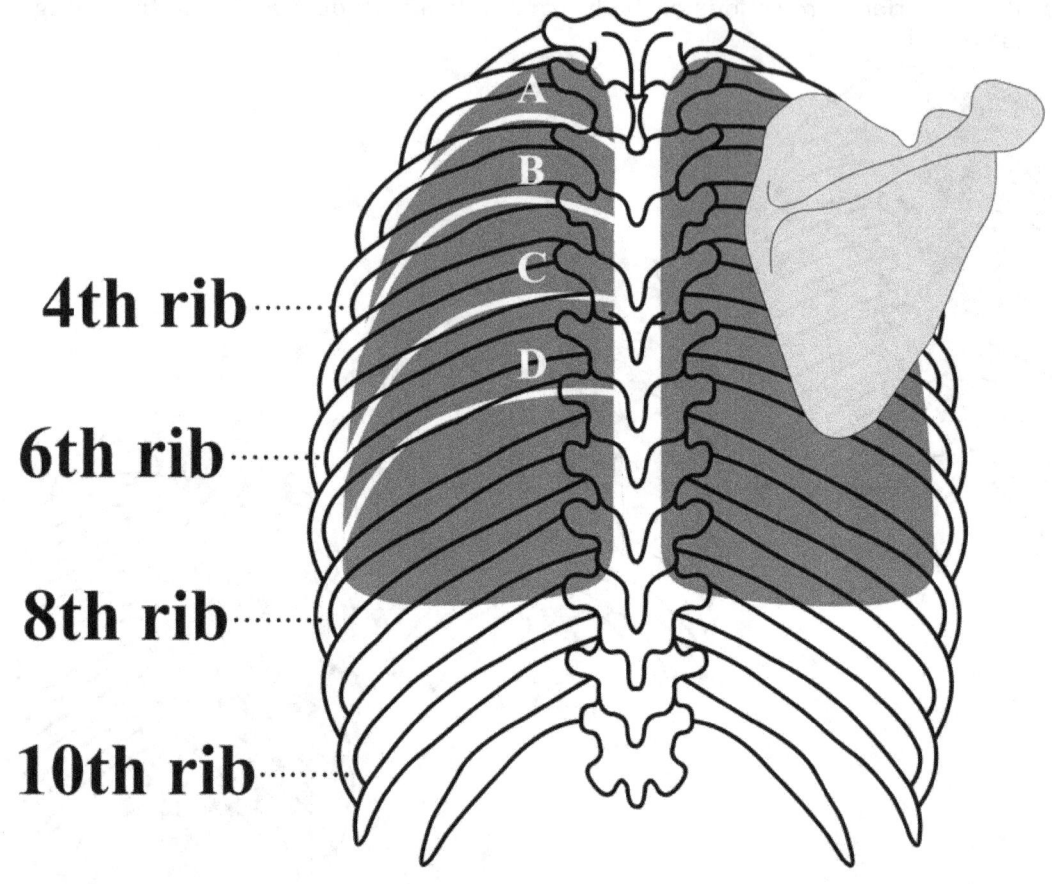

Fig. 9-7

37. Explain why aspirated material most commonly enters the right lung if a person is sitting, standing, lying supine, or lying on the right side and most commonly enters the left lung if a person is lying on the left side.

38. As each main stem bronchus enters the hilum of its lung, it gives rise to a lobar bronchus for each of the lung's lobes. In the right lung, the main stem bronchus divides into a laterally-directed right upper lobar bronchus and an inferolaterally directed intermediate bronchus which subsequently bifurcates into a laterally-directed right middle lobar bronchus and an inferiorly directed right lower lobar bronchus. In the left lung, the main stem bronchus divides into a laterally directed left upper lobar bronchus and an inferolaterally directed left lower lobar bronchus. Applying the reasoning discussed in the answer to the preceding question, identify the lobar bronchus that aspirated material most commonly enters when the person is in the following positions:

Sitting: _____
Standing: _____
Lying supine: _____
Lying on the right side: _____
Lying on the left side: _____

124

39. The lobar bronchi in each lung give rise to segmental bronchi. The segmental bronchi serve separate segments of the lung called bronchopulmonary segments. Discuss the shape and orientation of the bronchopulmonary segments within the lungs, the number of segmental bronchi that transmit air into each segment, and the pulmonary arterial supply and pulmonary venous drainage of the segments.

40. The depth of the costodiaphragmatic recess remains constant as it extends all around the costodiaphragmatic margin of each lung. At every point along the costodiaphragmatic margin of both lungs, the bottom of the costodiaphragmatic recess lies 2 rib levels below that of the costodiaphragmatic margin of the lung (given that the patient is breathing quietly). Based upon your answers to question 35, complete the following statement:
When a person is breathing slowly,
   the midclavicular line intersects the bottom of the costodiaphragmatic recess
      at the level of the _____ rib,
   the midaxillary line intersects the bottom of the costodiaphragmatic recess
      at the level of the _____ rib, and
   the lateral border of the spine intersects the bottom of the costodiaphragmatic recess
      at the level of the _____ rib.

Knowledge of the surface projections of each lung's apex and fissures and the rib levels which overlie its costodiaphragmatic margin and the bottom of its costodiaphragmatic recess must always be recalled when physically examining or performing a procedure on a patient's chest. The following 4 questions illustrate the importance of this knowledge:

41. Where do you place the stethoscope when you want to listen to the breath sounds in the apex of a patient's lung?
   _____ Immediately above the lateral third of the clavicle
   _____ Immediately above the medial third of the clavicle
   _____ Over the anterior end of the 2nd intercostal space
   _____ Over the intersection of the 2nd intercostal space with the midclavicular line

42. Cardiac failure (that is, the inability of the heart to maintain normal cardiac output) can lead to pleural effusion (accumulation of fluid within the pleural space). Pleural effusion that develops as a consequence of cardiac failure can sometimes be detected through auscultation or percussion of the lungs. With a patient suffering from cardiac failure who is sitting upright as you examine him/her, where would you expect almost all of the excess pleural fluid to be initially pooled in the pleural spaces of the lungs? Explain the anatomical basis of your answer.

43. When you place the stethoscope on a patient's right 4th intercostal space at a spot that intersects the right midclavicular line, you hear breath sounds in which lobe of the right lung?   _____ upper lobe   _____ middle lobe   _____ lower lobe

44. To perform a liver biopsy, a patient is first asked to lie in a supine position, that is, to lie on his or her back. The patient is asked to exhale as much as possible and then hold their breath as a needle is inserted into the chest at the 9th intercostal space along the midaxillary line. Explain why the needle will pass through the right lung's pleural space but not pierce the right lung's lower lobe.

45. Describe the sensory innervation of the visceral and parietal pleura.

Mediastinum

The mediastinum is the median region of the chest. It is bordered anteriorly by the sternum, posteriorly by the thoracic vertebrae, and, on each side, by mediastinal pleura. The mediastinum is bordered above by the thoracic inlet and below by the diaphragm.

The superior mediastinum and the inferior mediastinum are the parts of the mediastinum, respectively, above and below the level of the sternal angle.

46. The sternal angle lies at the level of the _____ thoracic intervertebral disc.

_____ 1st   _____ 2nd   _____ 3rd   _____ 4th   _____ 5th   _____ 6th   _____ 7th

The inferior mediastinum is divisible into three regions: The anterior mediastinum (which is the region of the inferior mediastinum between the sternum and the pericardium), the middle mediastinum (which is the mediastinal region consisting of the pericardium and its contents and the tracheal bifurcation), and the posterior mediastinum (which is the region of the inferior mediastinum between the pericardium and the spine).

Indicate the regions of the mediastinum either occupied or traversed by the following viscera:

Viscus	Superior Mediastinum	Anterior Mediastinum	Middle Mediastinum	Posterior Mediastinum
47. Arch of aorta	_____	_____	_____	_____
48. Ascending aorta	_____	_____	_____	_____
49. Azygos vein	_____	_____	_____	_____
50. Brachiocephalic trunk	_____	_____	_____	_____
51. Brachiocephalic veins	_____	_____	_____	_____
52. Carina	_____	_____	_____	_____
53. Descending thoracic aorta	_____	_____	_____	_____
54. Esophagus	_____	_____	_____	_____
55. Heart	_____	_____	_____	_____
56. Internal thoracic arteries	_____	_____	_____	_____
57. Left common carotid artery	_____	_____	_____	_____
58. Left recurrent laryngeal nerve	_____	_____	_____	_____
59. Left subclavian artery	_____	_____	_____	_____
60. Main stem bronchi	_____	_____	_____	_____
61. Phrenic nerves	_____	_____	_____	_____
62. Pulmonary arteries	_____	_____	_____	_____
63. Pulmonary trunk	_____	_____	_____	_____
64. Pulmonary veins	_____	_____	_____	_____
65. Superior vena cava	_____	_____	_____	_____
66. Thoracic duct	_____	_____	_____	_____
67. Thymus	_____	_____	_____	_____
68. Trachea	_____	_____	_____	_____
69. Vagus nerves	_____	_____	_____	_____

END OF QUESTIONS IN PART A OF THE CHAPTER ON THE CHEST (THORAX

CHEST (THORAX) – Part B: Questions and Answers

Dissection of the chest (thorax) in gross lab focuses on identification of
  (1) the sternum, ribs, costal cartilages, intercostal muscles, and intercostal arteries, veins, and nerves,
  (2) the heart's pericardial sac, surfaces, chambers, and valves,
  (3) the coronary arteries, cardiac veins, and coronary sinus,
  (4) the lobes of the lungs,
  (5) the trachea and the main stem and lobar bronchi of the lungs,
  (6) the branches of the ascending aorta, aortic arch, and descending thoracic aorta,
  (7) the brachiocephalic veins and the superior and inferior venae cavae,
  (8) the pulmonary trunk, ligamentum arteriosum, pulmonary arteries, and pulmonary veins,
  (9) the azygous vein and its tributaries,
  (10) the paired phrenic, paired vagus, and left recurrent laryngeal nerves, and
  (11) the esophagus, thoracic duct, and diaphragm.

The thoracic anatomy most frequently applied in clinical practice is the surface anatomy relevant to the physical examination of the heart and lungs.

Rib Cage

1. Describe the anatomical relationship in the anterior chest wall that permits quick identification of the ribs and the intercostal spaces in the anterior part of the rib cage.

   The most easily identifiable landmark in the anterior chest wall is the manubriosternal joint, the joint between the manubrium and the body of the sternum. The manubriosternal joint is a symphysis, that is, a secondary cartilaginous joint. The manubriosternal joint can be easily palpated as a rather prominent, horizontal ridge in the midline of the anterior chest wall. The joint's prominence is due in large part to the fact that the posterior angle between the manubrium and body of the sternum is less than 180°. This angle is called the sternal angle.

   The sternal angle is also the single most important landmark of the anterior chest wall. This is because, on each side, the costal cartilage of the 2nd rib almost invariably articulates with the sternum at the level of the sternal angle. Accordingly, palpation of the sternal angle permits unambiguous identification of not only the 2nd ribs at the sides of the sternal angle, but also the 3rd through 7th ribs (by simply counting them in their sequential palpation from the 2nd rib downward). Distinct identification of the 8th, 9th, and 10th ribs in the anterior chest wall is difficult because of their relatively close proximity to each other.

   The spaces between adjacent ribs are called intercostal spaces. Each intercostal space is identified by its number; the number corresponds to that of the rib bordering the upper margin of the space. For example, the intercostal space between the 5th and 6th ribs on the right side is called the right 5th intercostal space.

2. List all the ribs whose costal cartilages contribute to the costal margins of the rib cage.
   _____ 6th
   __x__ 7th
   __x__ 8th
   __x__ 9th
   __x__ 10th
   _____ 11th
   _____ 12th

127

3. Explain why the 1ˢᵗ ribs are not palpable in the anterior chest wall.

The 1ˢᵗ ribs are normally impalpable because the clavicles overlie them.

4. If a person is suffering from an abnormal collection of fluid or air in the pleural space around a lung, a chest tube may inserted through an intercostal space along the midaxillary line into the pleural space to remove the fluid or air. Explain why it is important to insert the chest tube **up and over the lower rib** bordering the intercostal space.

The intercostal nerve that extends along the length of each intercostal space lies partially under the cover of the costal groove of the upper rib bordering the intercostal space. Of the intercostal arteries and veins that extend along the length of each intercostal space, the largest intercostal arteries and veins extend alongside the intercostal nerve. Insertion of a chest tube up and over the lower rib bordering an intercostal space minimizes the risk of injury to the intercostal nerve and the largest intercostal arteries and veins in the intercostal space.

5. The nerve supply of the chest wall is provided by 11 pairs of intercostal nerves. Describe the derivation of the intercostal nerves from thoracic spinal nerves.

The 11 pairs of intercostal nerves are derived from the anterior rami of the first 11 thoracic spinal nerves and numbered accordingly. Each intercostal nerve, with the exception of the first one, begins as almost the entire anterior ramus of the corresponding thoracic spinal nerve. The first intercostal nerve differs from this general pattern in that it begins as merely a minor branch of the anterior ramus of T1. Most of the anterior ramus of T1 forms the T1 root of the brachial plexus.

6. The 1ˢᵗ through 9ᵗʰ intercostal spaces are each supplied by anterior and posterior intercostal arteries (the 10ᵗʰ and 11ᵗʰ intercostal spaces are each supplied by only a posterior intercostal artery). Describe the origins of the anterior and posterior intercostal arteries.

Whereas the anterior intercostal arteries of the 5 uppermost intercostal spaces are all branches of the internal thoracic artery, those of the next lower 4 intercostal spaces are all branches of the musculophrenic artery.

The posterior intercostal arteries of the 9 lowest intercostal spaces are all branches of the descending thoracic aorta. The posterior intercostal arteries of the two uppermost intercostal spaces are derived from an indirect branch of the subclavian artery.

7. Each intercostal space is drained by anterior and posterior intercostal veins. Describe the pathways by which blood drained by the anterior and posterior intercostal veins returns back to the right atrium of the heart.

Blood drained by anterior intercostal veins is conducted by the internal thoracic veins into the brachiocephalic veins and from the brachiocephalic veins into the right atrium of the heart via the superior vena cava. Blood drained by the posterior intercostal veins is conducted by the azygos system of veins into the azygos vein and from the azygos vein into the right atrium of the heart via the superior vena cava.

8. The prominence of the highest, readily palpable spinous process in the midline of the back can be enhanced by asking the patient to flex his/her head forward at the neck. This is the spinous process of the _____ vertebra.

_____ 4th cervical
_____ 5th cervical
_____ 6th cervical
__x__ 7th cervical
_____ 1st thoracic
_____ 2nd thoracic
_____ 3rd thoracic
_____ 4th thoracic

9. The 12th rib can be palpated in the back. <u>True</u> or False

10. In the back, the medial end of the spine of the scapula lies at the level of the spinous process of the _____ vertebra.

_____ 7th cervical
_____ 1st thoracic
_____ 2nd thoracic
__x__ 3rd thoracic
_____ 4th thoracic
_____ 5th thoracic
_____ 6th thoracic

11. In the back, the inferior angle of the scapula lies at the level of the _____ rib or its intercostal space.

_____ 5th
_____ 6th
__x__ 7th
_____ 8th
_____ 9th
_____ 10th

12. What is the common anatomical feature of thoracic vertebrae that results in the tip of the spinous process of each thoracic vertebra lying approximately at the level of the body of the vertebra below?

The thoracic vertebrae have the most inferiorly-directed spinous processes in the spine.

13. As a general rule, the anterior end of a rib lies at the level of the posterior end of the _____ rib below it.

_____ 1st
_____ 2nd
_____ 3rd
__x__ 4th
_____ 5th
_____ 6th
_____ 7th

14. Fig. 9-1 presents 12 different surface projections of the diaphragm's domes and central tendon to the anterior chest wall. Select the drawing which most accurately represents the anterior surface projection of the diaphragm's domes and central tendon in a healthy patient lying supine on an examination table and breathing quietly.

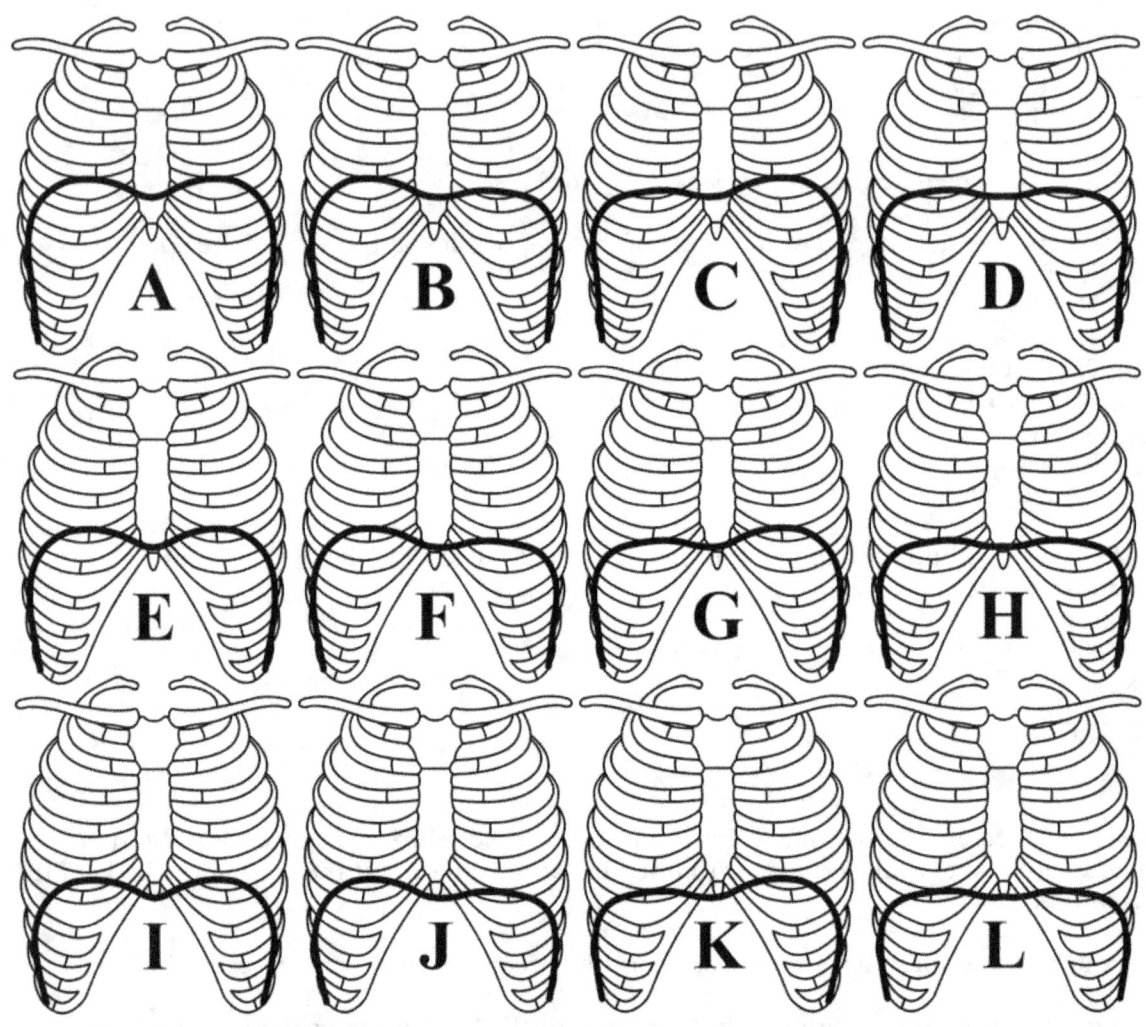

Fig. 9-1

The drawings in the first and fourth columns (drawings A-E-I and D-H-L) are not correct because the left and right domes of the diaphragm do not lie at the same level. The drawings in the third column (drawings C-G-K) are not correct because the right dome typically lies at a higher level than that of the left dome. Whereas the diaphragm's right dome directly overlies the liver's right lobe, the left dome directly overlies the fundus of the stomach. The fundus of the stomach typically lies at a level below that of the liver's right lobe. Of the drawings in the second column (drawings B-F-J), drawing F most accurately represents the anterior surface projection of the diaphragm's domes and central tendon in a healthy adult patient lying supine on an examination table and breathing quietly. Under these conditions, the top of the right dome intersects the right midclavicular line at about the level of the upper border of the right 5th rib, and the top of the left dome intersects the left midclavicular line at about the level of the lower border of the left 5th rib.

15. There are three major openings in the diaphragm for transmission of structures between the thorax and abdomen. Each opening is named for the largest structure which passes through it. Indicate the vertebral level at which the aortic (A), caval (C), and esophageal (E) openings lie when a person is seated upright and breathing quietly.

_____ body of 6th thoracic vertebra
_____ body of 7th thoracic vertebra
_C_ body of 8th thoracic vertebra
_____ body of 9th thoracic vertebra
_E_ body of 10th thoracic vertebra
_____ body of 11th thoracic vertebra
_A_ body of 12th thoracic vertebra
_____ body of 1st lumbar vertebra
_____ body of 2nd lumbar vertebra

16. The aortic (A), caval (C), esophageal (E) openings of the diaphragm each transmit two or more structures. Identify the opening through which each of the following structures passes:

_A_ Azygos vein
_E_ Branches of the left gastric artery
_E_ Left and right vagus nerves
_C_ Right phrenic nerve
_A_ Thoracic duct
_E_ Tributaries of the left gastric vein

17. Which nerves innervate the diaphragm muscle?

The left phrenic nerve innervates the left side of the diaphragm muscle, and the right phrenic nerve innervates the right side of the diaphragm muscle.

18. Which spinal nerves contribute nerve fibers to the phrenic nerve?

_____ C1
_____ C2
_x_ C3
_x_ C4
_x_ C5
_____ C6
_____ C7
_____ C8

19. Fig. 9-2 represents the anterior chest wall of an adult male. The locations of the most prominent surfaces features are labelled: C's for the upper borders of the clavicles, JN for the jugular notch, SA for the sternal angle, N's for the nipples, XP for the tip of the xiphoid process, and CM's for the costal margins of the rib cage. Each nipple is located in the 4th intercostal space immediately lateral to the midclavicular line. In the appropriate squares, print an A for the site where the closure of the aortic valve can be best heard, a P for the site where the closure of the pulmonary valve can be best heard, an M where the closure of the mitral valve can be best heard, and a T where the closure of the tricuspid valve can be best heard.

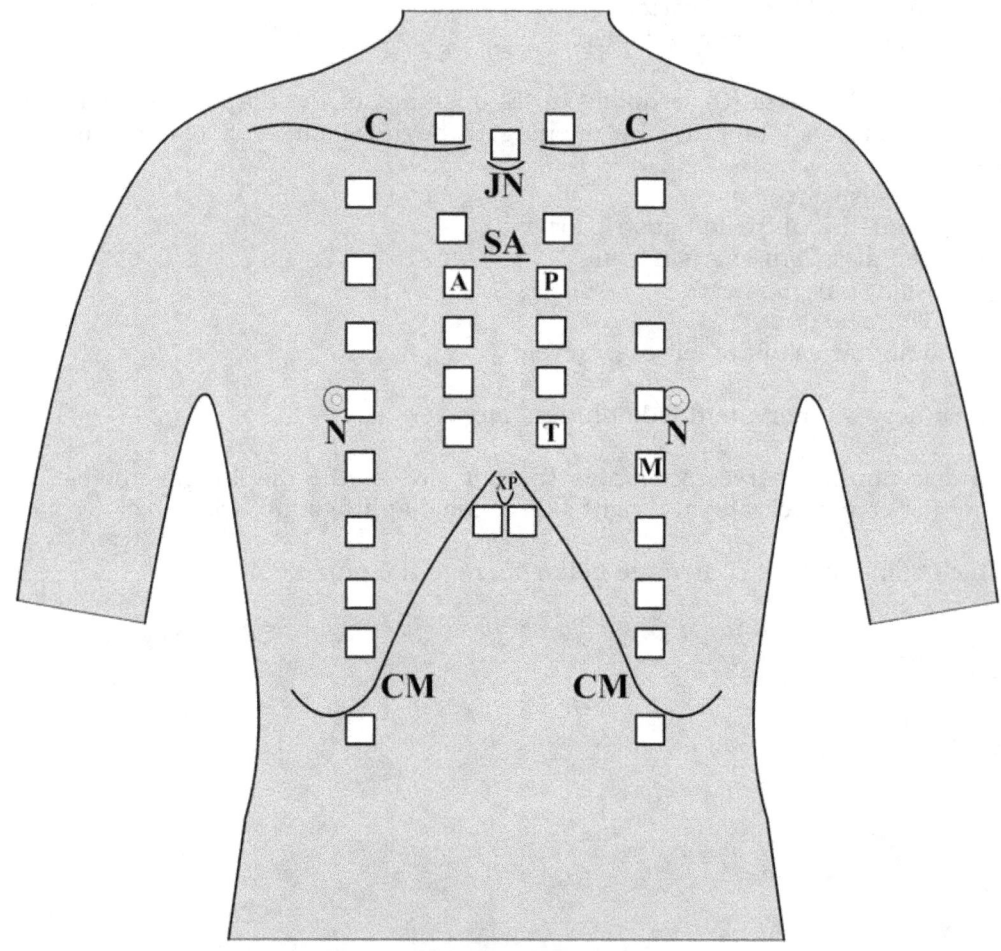

Fig. 9-2

The sites for best hearing the closure of the heart's valves are as follows:
Aortic valve – anterior end of right 2nd intercostal space
Pulmonary valve – anterior end of left 2nd intercostal space
Tricuspid valve – anterior end of left 5th intercostal space
Mitral valve – left 5th intercostal space at the intersection with the left midclavicular line

20. Fig. 9-3 presents 9 different heart locations in an adult with a normal sized heart. As in Fig. 9-2 on the preceding page, each nipple is located in the 4th intercostal space immediately lateral to the midclavicular line. Select the drawing with the most accurate heart location in an adult patient with a normal sized heart who is seated upright and breathing quietly as he/she is being examined.

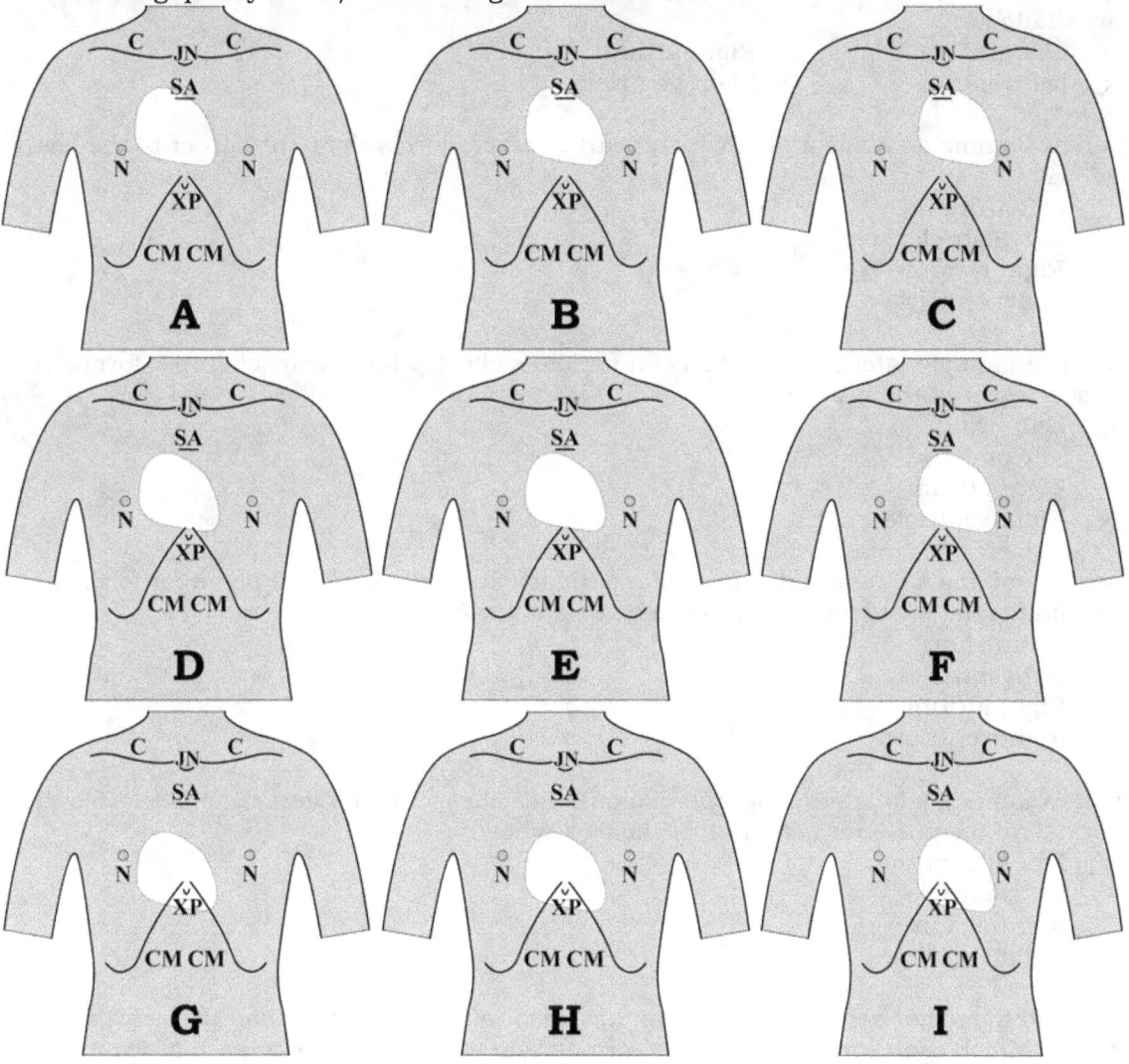

Fig. 9-3

Drawing F shows the most accurate heart location. Drawings A, B, D, E, G, and H are incorrect because about two-thirds of the heart lies to the left of the midline. Drawings C and I are not as accurate as drawing F because drawing C places the apex of the heart too high and drawing I places the apex of the heart too low. In an adult patient with a normal sized heart who is seated upright and breathing quietly as he/she is being examined, the apex of the heart lies in the left 5th intercostal space near the intersection with the left midclavicular line. The diaphragmatic surface of the heart and the underlying central tendon of the diaphragm lie at about the level of the xiphisternal joint (the joint between the body of the sternum and the xiphoid process). The uppermost extent of the heart lies at about the level of the sternal angle. The right border of the heart lies about an inch to the right border of the body of the sternum. Relative to the rib cage, the right border of the heart extends from the right 3rd costal cartilage to about the right 6th costal cartilage. The left border of the heart extends from the anterior end of the left 2nd intercostal space to the apex of the heart.

21. Which heart chamber forms the apex of the heart?

_____ Left atrium   _____ Right atrium
__x__ Left ventricle   _____ Right ventricle

22. When examining a PA chest film, which heart chamber forms most of the left border of the heart shadow?

_____ Left atrium   _____ Right atrium
__x__ Left ventricle   _____ Right ventricle

23. When examining a PA chest film, which heart chamber forms the right border of the heart shadow?

_____ Left atrium
_____ Left ventricle
__x__ Right atrium
_____ Right ventricle

24. When examining a lateral radiograph of a patient's chest, which heart chamber forms the anterior border of the heart shadow?

_____ Left atrium
_____ Left ventricle
_____ Right atrium
__x__ Right ventricle

25. When examining a lateral radiograph of a patient's chest, which heart chamber forms the upper half of the posterior border of the heart shadow?

__x__ Left atrium
_____ Left ventricle
_____ Right atrium
_____ Right ventricle

26. When examining a lateral radiograph of a patient's chest, which heart chamber forms the lower half of the posterior border of the heart shadow?

_____ Left atrium
__x__ Left ventricle
_____ Right atrium
_____ Right ventricle

27. Almost all the major coronary arteries and cardiac veins lie in grooves on the heart's surface. Match each coronary artery or cardiac vein with the groove or path it occupies.
A: Anterior interventricular groove
B: Inferior interventricular groove
C: Atrioventricular groove on heart's sternocostal surface
D: Atrioventricular groove on heart's posteroinferior surface
E: Short path leading to point where the anterior interventricular groove meets the atrioventricular groove on the heart's sternocostal surface

__A__ Left anterior descending artery (LAD)   __A__ Great cardiac vein
__B__ Posterior descending artery (PDA)   __B__ Middle cardiac vein
__E__ Left coronary artery   __D__ Coronary sinus
__C__ Right coronary artery

28. Which coronary artery generally supplies both the SA and AV nodes of the heart's conducting system?

_____ LAD   _____ PDA   __x__ Right coronary artery

134

29. Fig. 9-4 represents the anterior chest wall of an adult male. The locations of the most prominent surfaces features are labelled: C's for the upper borders of the clavicles, JN for the jugular notch, SA for the sternal angle, N's for the nipples, XP for the tip of the xiphoid process, and CM's for the costal margins of the rib cage. Each nipple is located in the 4th intercostal space immediately lateral to the midclavicular line. If this drawing represented the chest of an adult male who has just suffered a heart attack, mark the box with an "X" where you would press down on the chest to apply CPR (cardiopulmonary resuscitation). Explain the anatomical basis of your decision.

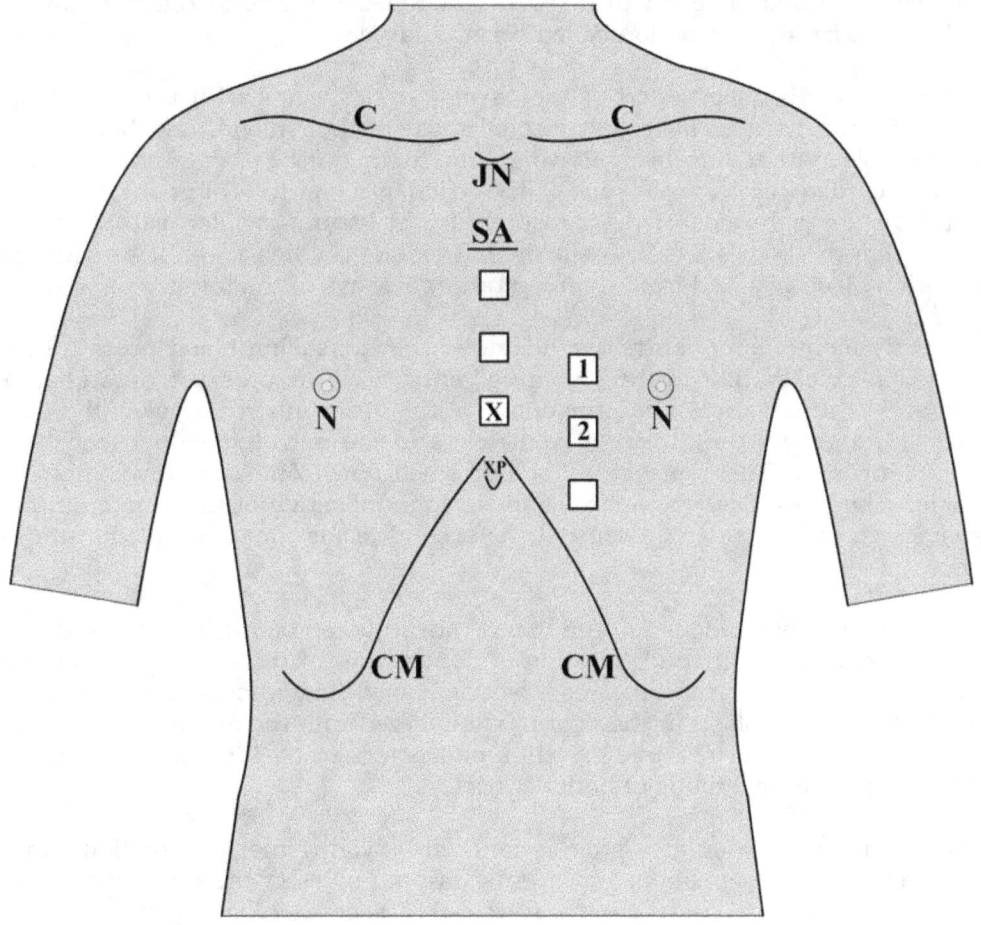

Fig. 9-4

Upon answering question 20 in this chapter, a person might believe that the best location for applying CPR would be either at the box marked 1 or the box marked 2. It might seem reasonable that these locations are sites where effective pressure can be applied to the heart's ventricles, as each of these locations overlie much of the heart's ventricles. However, neither of these sites is optimal for administration of CPR because when you apply pressure to the anterior chest wall during CPR, you want to press the heart against an unyielding structure behind the heart, and the unyielding structure that is closest to the back of the heart is the spine. Consequently, you want to press down on the heart in the midline of the chest. Every time you press down on the lower half of the body of the sternum during CPR, you are very effectively squeezing the heart's ventricles between the sternum and the spine. Under optimal conditions, 100 compressions per minute can generate about 30% of the heart's resting cardiac output.

30. Describe blood flow through the heart and the timing of the opening and closing of the heart's valves during the cardiac cycle.

On each side of the heart, the atrium and ventricle work together as an atrioventricular pump. The electrical activity conducted through the heart during each heartbeat synchronizes the pumping activities of the two atrioventricular pumps. The synchronous pumping activities of the two atrioventricular pumps can be described by a single cycle of events called the cardiac cycle. The cardiac cycle is divided into two periods: the cycle begins with a period of ventricular relaxation called ventricular diastole, and ends with a period of ventricular contraction called ventricular systole.

The ventricular diastolic period of the cardiac cycle begins with the heart's musculature relaxed. Both ventricles have just ejected most of their blood, and both atria are nearly filled with blood that has been flowing into them during the terminal moments of the preceding cardiac cycle. All valves (the tricuspid, mitral, pulmonary, and aortic) are closed. On the right side of the heart, the right atrium is filled with oxygen-poor blood received from the body's tissues. On the left side of the heart, the left atrium is filled with oxygen-rich blood received from the respiratory airways of the lungs.

In the early moments of ventricular diastole, the increasing blood pressure in each filled atrium soon exceeds that in the adjoining ventricle. This pressure difference between the atrium and ventricle forces open the cusps of the atrioventricular valve (the tricuspid valve between the right atrium and right ventricle and the mitral valve between the left atrium and left ventricle). There ensues a flood of blood flow from each atrium into its adjoining ventricle. The blood flow soon ebbs, and throughout much of the remainder of ventricular diastole, each atrium serves essentially as a conduit for conveying blood into its adjoining ventricle.

Blood flow into both sides of the heart abruptly ends during the final moments of ventricular diastole with the beginning of contraction of the cardiac muscle tissue in the walls of both atria. During atrial contraction, blood flow into each atrium is momentarily blocked as all of the blood in the atrium is squeezed into the adjoining ventricle. The final moments of ventricular diastole are thus characterized by the near-complete emptying of both atria and the final filling of both ventricles.

The beginning of ventricular systole is marked by ventricular contraction concurring with atrial relaxation. These concurrent events on both sides of the heart promptly generate a blood pressure in each contracting ventricle that exceeds blood pressure in the adjoining, relaxing atrium. This pressure difference across each atrioventricular valve closes the valve shut. The tricuspid and mitral valves close shut almost simultaneously, producing immediately after their closure vibrations in the ventricular walls and in the blood confined to the ventricular chambers. When these vibrations reach the chest wall, they collectively form S1, the first heart sound (the "lub" sound) of each heartbeat.

Continuing ventricular contraction now markedly elevates the blood pressure in each ventricle until it distends the root of the arterial trunk emanating from the ventricle (the pulmonary trunk emanating from the right ventricle and the aorta emanating from the left ventricle). There occurs a sudden ejection of the high-pressured, ventricular blood into the artery at the moment that the apposed cusps of the valve at the artery's origin become separated by the artery's distention.

The relaxation of the atrial musculature and the closure of the atrioventricular valves during the early moments of ventricular systole initiate the filling of both atria again. The atria serve as reservoirs of blood until the atrioventricular valves reopen during the early moments of the ventricular diastolic period of the next cardiac cycle.

Toward the end of ventricular systole, after the ventricles have ejected most of their blood, there begins a backward flow of blood from the pulmonary trunk and aorta into the ventricles (as a consequence of the blood in each arterial trunk being greater than that in the ventricle). This backward flow snaps the pulmonary and aortic valves shut. The pulmonary and aortic valves snap shut almost simultaneously, generating immediately after their closure vibrations in the walls of both arterial trunks and in the blood they bear. When these vibrations reach the chest wall, they collectively produce S2, the second heart sound (the "dup" sound) of the heartbeat. At this time, the cardiac cycle begins anew.

Lungs and Pleura

The lungs occupy the left and right lateral regions of the chest. A microscopically-thin, fluid-filled space, called a pleural space, envelops each lung. The pleural space provides a lubricated free space between the lung and the chest wall regions which the lung faces. The pleural space also partially separates the lung from the thoracic viscera which lie between the lungs in the mid-region of the chest (the mid-region of the chest is called the mediastinum).

Each lung is basically pyramid-shaped, and has an apex and three external surfaces. The three surfaces are identically named in both lungs: each lung has a mediastinal surface which faces mediastinal viscera, a diaphragmatic surface which lies atop the diaphragm, and a costal surface which is apposed to the inner surface of the rib cage.

Each lung is attached to mediastinal viscera by a bundle of structures called the root of the lung. The root of each lung consists of all those structures which enter or exit the lung; the most prominent structures are the lung's main stem bronchus, pulmonary artery, and pulmonary veins. The hilum of each lung is that region of the lung's mediastinal surface at which structures are transmitted between the root of the lung and the parenchyma of the lung (the parenchyma of an organ consists of the organ's elements that are essential to its functions).

31. Describe the parenchymal elements of the lungs and their functions.

The total aggregate of airways in a lung is known as its bronchial tree. There are two basic types of airways in the bronchial tree: conduction airways and respiratory airways. The conduction airways are the largest airways of the bronchial tree; they serve only as conduits for the mass flow of air between the trachea and the respiratory airways. The respiratory airways are the smallest airways of the bronchial tree; they provide diffusional exchange of oxygen and carbon dioxide between (a) the air breathed into the lungs and (b) the blood in the pulmonary circulation.

The pulmonary artery of each lung conducts oxygen-poor blood from the right side of the heart to the lung for gaseous exchange. The pulmonary artery gives rise to an arterial vascular tree in the lung whose branching pattern matches, in general, that of the bronchial tree. The smallest arterial branches conduct oxygen-poor blood into the capillary networks which envelop the respiratory airways, and it is here that the blood becomes oxygenated.

The capillary networks which envelop the respiratory airways are drained by minute venules that conduct the oxygen-rich blood toward the lung's pulmonary veins. The pulmonary veins conduct the oxygen-rich blood to the left side of the heart for distribution throughout the body.

32. Describe the mechanics of respiration. In particular, discuss the forces which expand the lungs when you breathe in and the forces which retract the lungs when you breathe out.

Respiration, or breathing, consists of two phases: inspiration, or breathing in, and expiration, or breathing out. Respiration can be classified as quiet or forced. The relaxed breathing which occurs when resting or sleeping is called quiet respiration. The strenuous breathing which occurs when exercising vigorously is called forced respiration.

Inspiration is driven by movements of the diaphragm and rib cage. The diaphragm is the chief muscle of inspiration. Contraction of the diaphragm during inspiration lowers the diaphragm within the trunk of the body. Descent of the diaphragm within the trunk of the body accounts for almost all inspiration during quiet respiration. During forced respiration, descent of the diaphragm within the trunk of the body is accompanied by an outward, anterior movement of the sternum and the anterior ends of the upper ribs and an outward lateral movement of the lower ribs. The diaphragmatic and costal surfaces of the lungs follow these inspiratory movements of the diaphragm and rib cage because, under normal physiological conditions, the pleural fluid in the pleural space around each lung has a pressure which averages 8 to 10 mm Hg below atmospheric pressure. The pleural fluid's subatmospheric pressure, in effect, presses each lung's diaphragmatic surface against the upper surface of the diaphragm and the lung's costal surface against the inner surface of the rib cage. The pleural fluid's subatmospheric pressure thus ensures that the lung's costal and diaphragmatic surfaces always follow any inspiratory movements of the diaphragm and rib cage.

The descent of the diaphragm and the outward movements of the rib cage expand the lungs. As each lung expands, the aggregate volume of its bronchial tree increases, and thus the average pressure within its airways decreases. When this average inner airway pressure becomes less than atmospheric pressure, air flows into the lung's bronchial tree.

Expiration is driven by the lungs' own internal, retractive forces. There are two sources of the lungs' retractive forces. (1) A retractive tension is exerted throughout expiration by each lung's elastic connective tissue fibers; these fibers were stretched during inspiration. (2) Most of each lung's volumetric increase during inspiration occurs through alveolar expansion. The expansion of the alveoli increases the surface tension of the surfactant lining the alveoli. Surfactant is a lipoprotein mixture secreted by type II alveolar epithelial cells. It forms a multilayered film at the interface between the fluid lining the luminal surface of the alveoli and the air in the alveoli. Surfactant is required for normal alveolar expansion during inspiration and retraction during expiration. The surface tension exerted by surfactant during expiration accounts for about two-thirds of the lung's retractive forces.

When the diaphragm and other inspiratory muscles relax during expiration, the lung's retractive forces effect a smooth recoil contraction of the lung. The diaphragm follows the upward movement of the lungs' diaphragmatic surfaces and the sternum and ribs follow the inward movements of the lungs' costal surfaces. As each lung retracts, the aggregate volume of its bronchial tree decreases, and thus the average pressure within its airways increases. When the average inner airway pressure becomes greater than atmospheric pressure, air flows out from the lung's bronchial tree.

33. When conducting a physical examination of a patient or performing a clinical procedure anatomically related to the lungs and their pleural spaces (such as inserting a chest tube or inserting a central line into the subclavian vein), it is important to know the bony and cartilaginous structures of the rib cage which mark the surface projections of the lungs, their lobes, and pleural spaces. Fig. 9-5 presents drawings of 16 different surface projections of the lungs (not the pleural spaces) to the anterior chest wall. Select the drawing which most accurately represents the anterior surface projections of the lungs in a healthy adult patient sitting upright and breathing quietly.

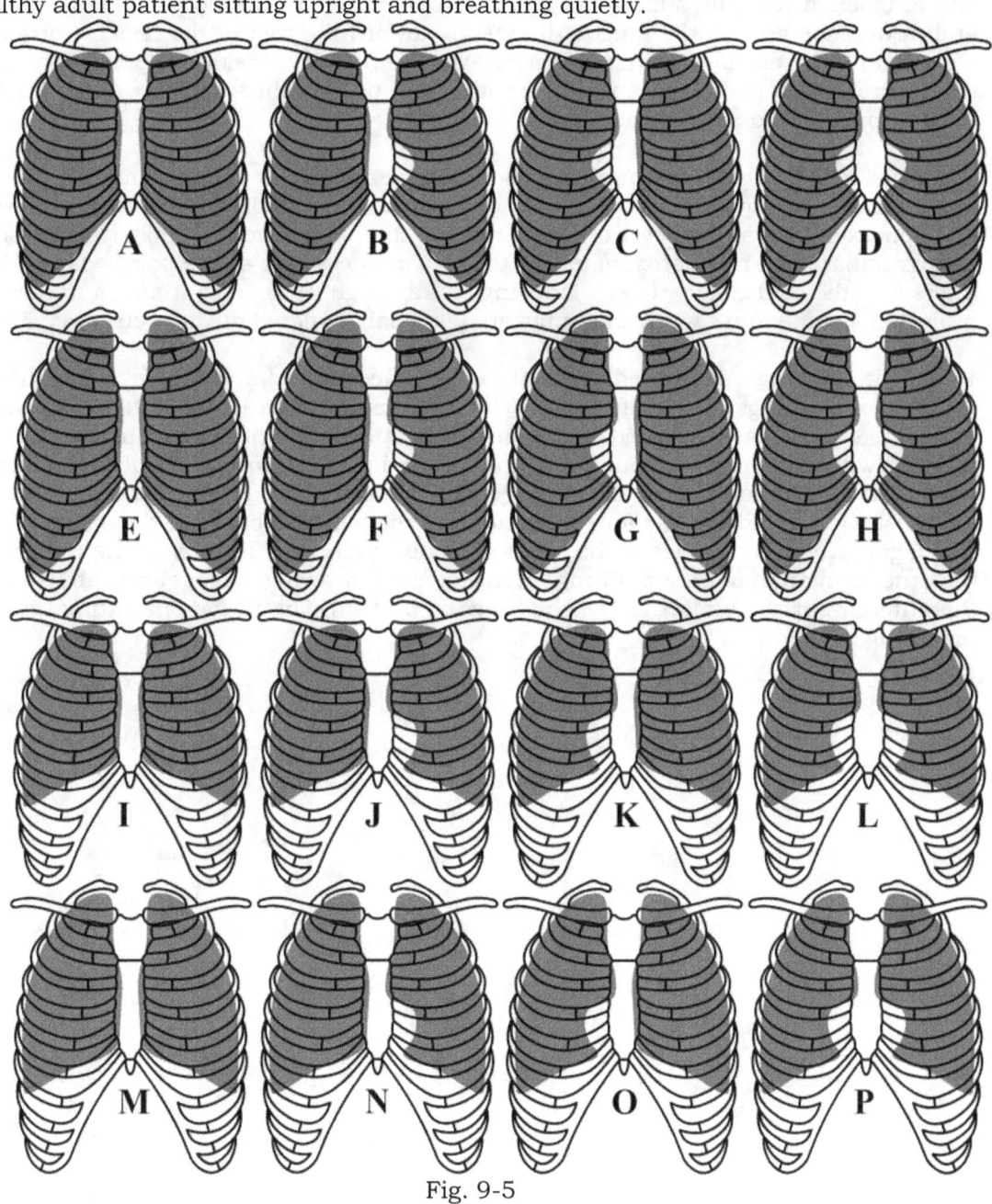

Fig. 9-5

139

Drawings A through D and I through L are not correct because the apices of the lungs do not project above the medial thirds of the clavicles. In an adult, the apex of each lung projects 2-4 cm above the medial third of the clavicle. You can hear vesicular breath sounds emanating from the apex of each lung when you place a stethoscope immediately above the medial third of the clavicle. Vesicular breath sounds are fine, low-pitched sounds heard throughout inspiration and early expiration. The projection of the apex of each lung above the medial third of the clavicle must also be recalled any time a central line is inserted into the subclavian vein. On each side of the body, the subclavian vein arches over the apex of the lung as it extends toward the root of the neck. Consequently, placement of a central line into the subclavian vein always risks the possibility that that the pleural space above the apex of the lung will be breached and the patient will suffer either a pneumothorax or hemothorax.

Drawings E through H are not correct because the lungs are not maximally inflated when a person breathes quietly. When a person breathes quietly, the lower margin of each lung lies approximately at the level of the 6th rib at the point where the 6th rib intersects the midclavicular line and approximately at the level of the 8th rib at the point where the 8th rib intersects the midaxillary line. Drawings E through H represent the anterior surface projections of the lungs when the lungs are maximally inflated during active respiration.

Now that we have established the anterior surface projections of the apices and lower margins of the lungs, we can focus on which of the drawings in the bottom row (drawings M to P) is correct. As discussed in the answer to question 20, the heart has an asymmetric position within the chest: about two-thirds of the heart lies to the left of the midline. The projection of the heart's apex all the way to the midclavicular line at the level of the left 5th intercostal space indents the anterior margin of the left lung between the 4th and 6th costal cartilages. The right border of the heart does not project sufficiently to the right to indent the anterior margin of the right lung. Consequently, drawing N most accurately represents the anterior surface projections of the lungs in a healthy adult patient sitting upright and breathing quietly.

34. Fig. 9-6 presents drawings of 3 different surface projections of the lungs (not the pleural spaces) to the posterior chest wall. Select the drawing which most accurately represents the posterior surface projections of the lungs in a healthy adult patient sitting upright and breathing quietly.

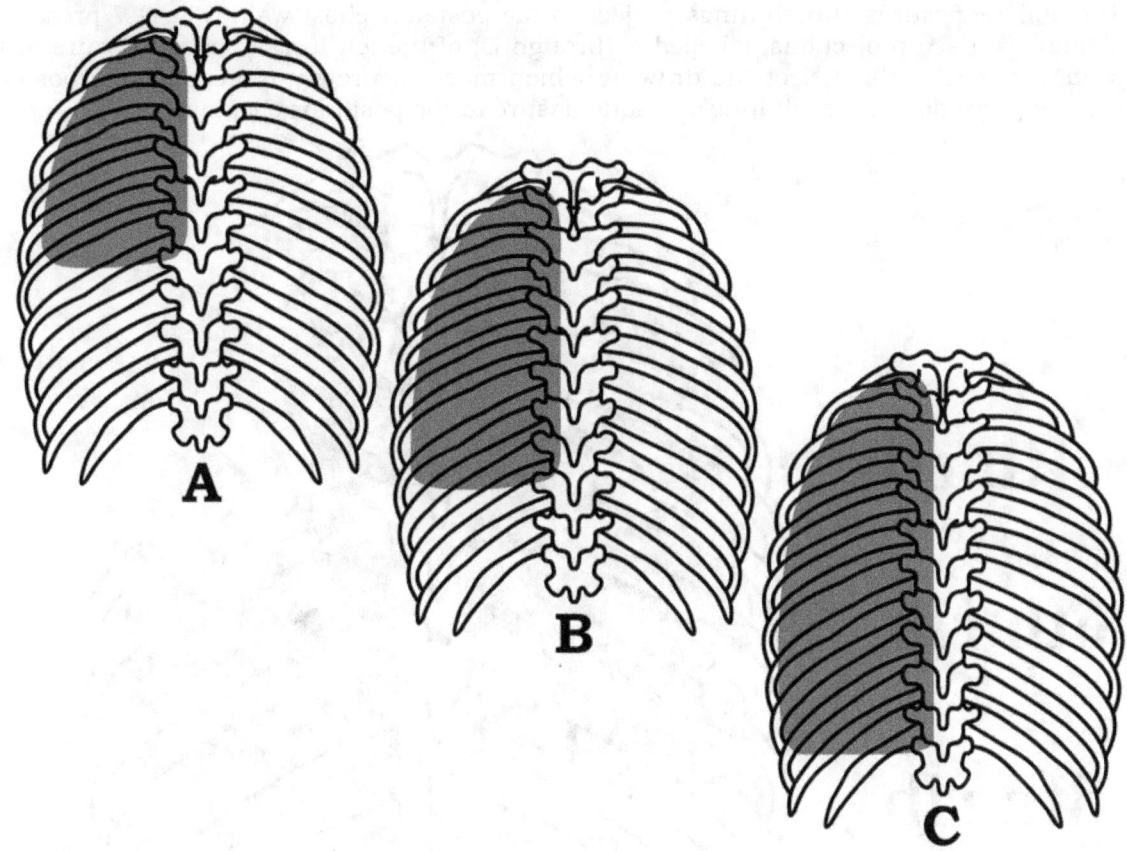

Fig. 9-6

Drawing B most accurately represents the posterior surface projections of the lungs. When a person breathes quietly, the lower margin of each lung lies approximately at the level of the 10th rib at the point where the 10th rib intersects the lateral border of the spine.

35. Based upon your answers to questions 33 and 34, complete the following statement:
When a person is seated upright and breathing slowly,
the midclavicular line intersects the lower margin of either lung
at the level of the __6th__ rib,
the midaxillary line intersects the lower margin of the lung
at the level of the __8th__ rib, and
the lateral border of the spine intersects the lower margin of the lung
at the level of the __10th__ rib.

Each lung has a deep fissure, called the oblique fissure, which runs obliquely through the lung. The right lung only has an additional fissure, called the horizontal fissure, which extends horizontally through it.

The two fissures of the right lung divide it into three lobes: an upper lobe, a middle lobe, and a lower lobe. The upper lobe of the right lung is the lobe which lies above both fissures, the middle lobe is the lobe which lies between the two fissures, and the lower lobe is the lobe which lies below the oblique fissure.

141

The single fissure of the left lung divides it into two lobes: an upper lobe and a lower lobe. The anteroinferior region of the upper lobe is called the lingula; the lingula is considered to be the left lung's homologue of the right lung's middle lobe.

36. The oblique fissures of both lungs project to the posterior chest wall. Fig. 9-7 presents 4 different surface projections, labelled A through D, of the left lung's oblique fissure to the posterior chest wall. Select the drawing which most accurately represents the posterior surface projection of the left lung's oblique fissure to the posterior chest wall.

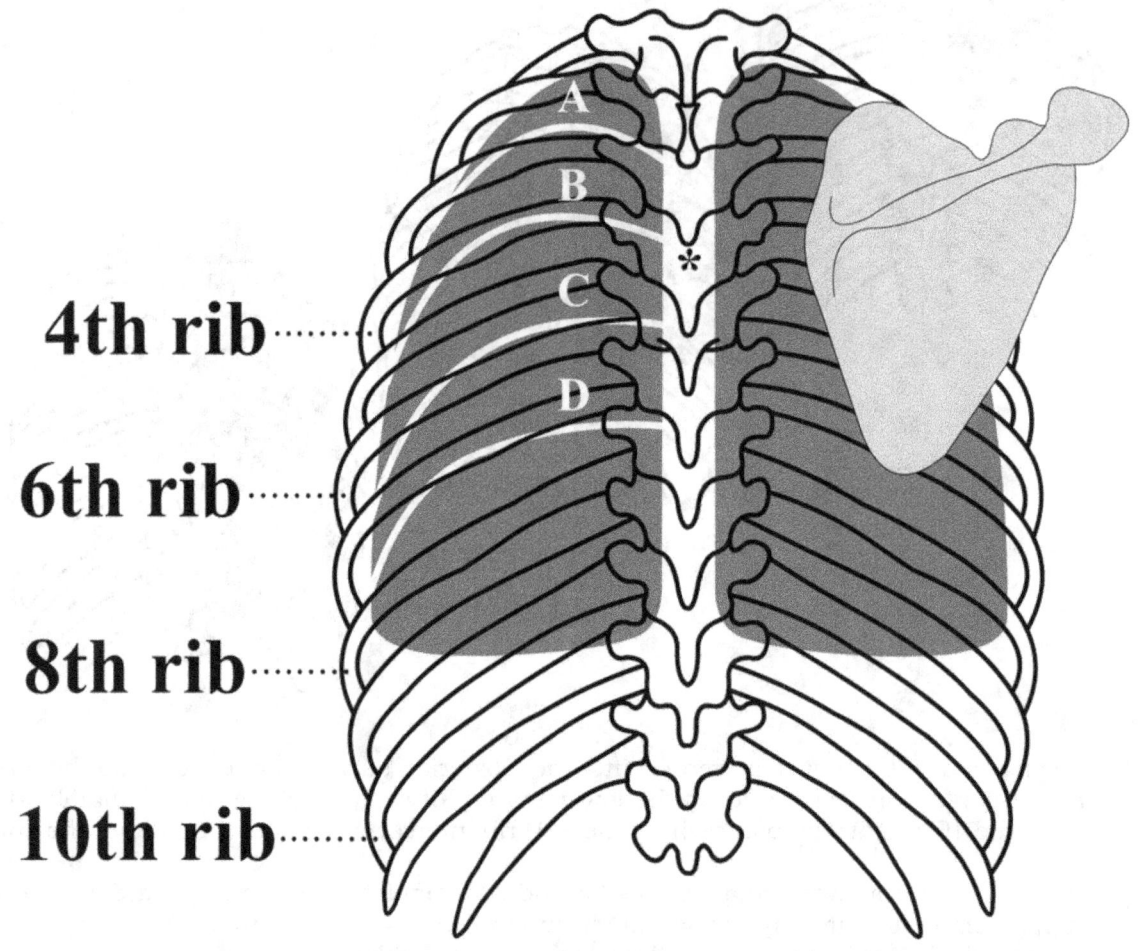

Fig. 9-7

Surface projection B is the most accurate projection. For both lungs, the surface projection of the oblique fissure to the posterior chest wall is an upward arching arc that begins near the level of the spinous process of the 3rd thoracic vertebra and ends laterally at about the intersection of the 5th intercostal space with the midaxillary line. You may recall from question 10 in this chapter that the medial end of the spine of the scapula lies at the level of the spinous process of the 3rd thoracic vertebra (the asterisk lies immediately below the spinous process of the 3rd thoracic vertebra). If a patient raises both arms and places both hands behind the head, the medial borders of the laterally rotated scapulae approximate the surface projections of the oblique fissures to the back. When listening to breath sounds or percussing the lungs on the posterior chest wall of a patient, keep in mind that each lung's upper lobe lies deep to about only the upper third or fourth of the posterior chest wall; the lower lobe lies deep to the remainder of the posterior chest wall, down to about the level of the posterior end of the 10th rib.

Fig. 9-8 shows that, for both lungs, the surface projection of the oblique fissure (OF) to the anterior chest wall is a laterally arching arc that begins laterally at about the level of the 5th intercostal space and ends medially at the intersection of the 6th rib with the midclavicular line. For the right lung, the horizontal fissure (HF) follows the course of the 4th rib from the midaxillary line to the lateral border of the sternum. Each lung's upper lobe projects more surface area to the anterior chest wall than its lower lobe.

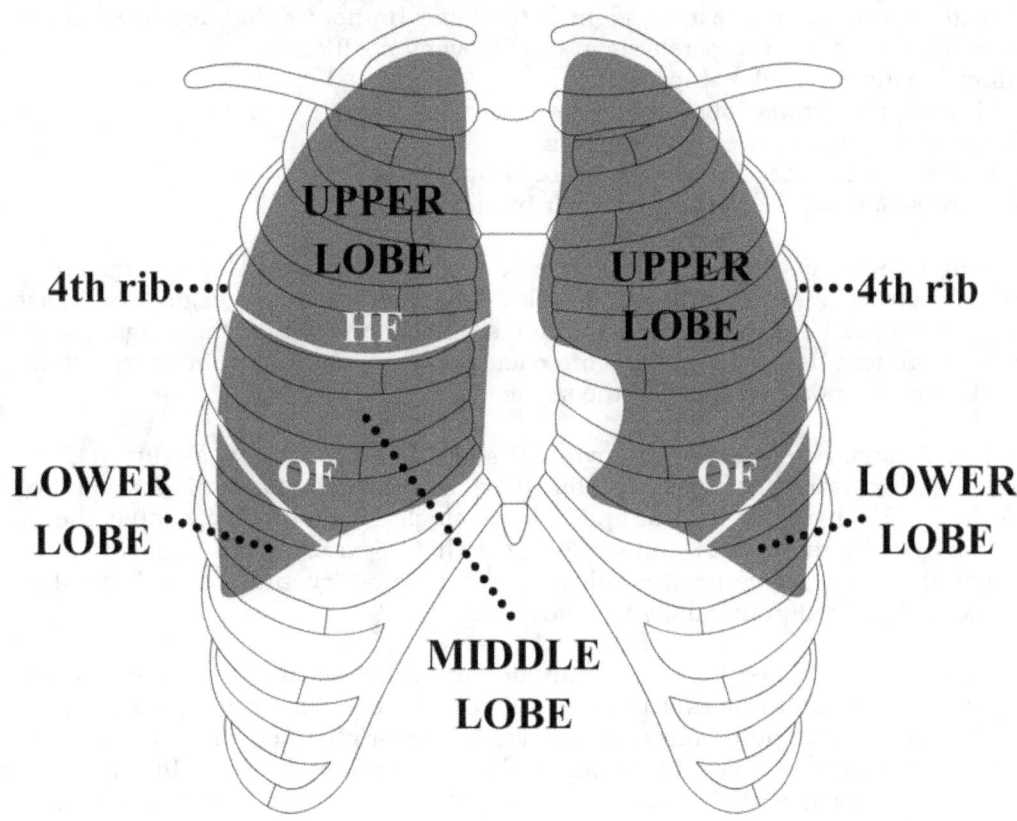

Fig. 9-8

37. Explain why aspirated material most commonly enters the right lung if a person is sitting, standing, lying supine, or lying on the right side and most commonly enters the left lung if a person is lying on the left side.

The trachea bifurcates into the left and right main stem, or primary, bronchi near or at the level of the sternal angle. The right main stem bronchus is more vertical, larger in diameter, and shorter in length than the left main stem bronchus. Consequently, aspirated material most commonly enters the right lung if a person is sitting, standing, lying supine, or lying on the right side because, in each of these positions, the right main stem bronchus is either the more direct continuation of the trachea or more aligned with the earth's gravitational field than the left main stem bronchus. Aspirated material most commonly enters the left lung if a person is lying on the left side because, in this position, the left main stem bronchus is more aligned with the earth's gravitational field than the right main stem bronchus.

38. As each main stem bronchus enters the hilum of its lung, it gives rise to a lobar bronchus for each of the lung's lobes. In the right lung, the main stem bronchus divides into a laterally-directed right upper lobar bronchus and an inferolaterally directed intermediate bronchus which subsequently bifurcates into a laterally-directed right middle lobar bronchus and an inferiorly directed right lower lobar bronchus. In the left lung, the main stem bronchus divides into a laterally directed left upper lobar bronchus and an inferolaterally directed left lower lobar bronchus. Applying the reasoning discussed in the answer to the preceding question, identify the lobar bronchus that aspirated material most commonly enters when the person is in the following positions:

 Sitting: __right lower lobar bronchus__

 Standing: __right lower lobar bronchus__

 Lying supine: __right lower lobar bronchus__

 Lying on the right side: __right upper lobar bronchus__

 Lying on the left side: __left upper lobar bronchus__

39. The lobar bronchi in each lung give rise to segmental bronchi. The segmental bronchi serve separate segments of the lung called bronchopulmonary segments. Discuss the shape and orientation of the bronchopulmonary segments within the lungs, the number of segmental bronchi that transmit air into each segment, and the pulmonary arterial supply and pulmonary venous drainage of the segments.

Bronchopulmonary segments are pyramid-shaped; their apices all project toward the hilum of the lung, and their bases all together form the costal and diaphragmatic surfaces of the lung. Each segment is supplied by a single segmental bronchus; the segmental bronchus conducts almost all the air which enters and exits the segment. Bronchopulmonary segments are the smallest respiratory units of a lung that can be identified and excised individually by a surgeon.

Each segment is supplied by a single pulmonary arterial branch. In other words, all the oxygen-poor blood that enters a bronchopulmonary segment for gaseous exchange enters via a single segmental pulmonary artery. The oxygen-rich blood that exits a segment exits partly via intrasegmental veins and partly via intersegmental veins. Intrasegmental veins extend along a segment's airways; by contrast, intersegmental veins course in the connective tissue septa separating adjacent segments. Whereas intrasegmental veins drain oxygen-rich blood from only their own segment, intersegmental veins receive oxygen-rich blood from at least two adjacent segments.

Therefore, there is a marked difference between the segmental distribution of the pulmonary arterial branches and the segmental distribution of the tributaries of the pulmonary veins in each lung. Pulmonary arterial branches form just one arborescent vascular system in a bronchopulmonary segment, one which parallels that of the airways. By contrast, pulmonary venous tributaries form two arborescent vascular systems among the segments, one which parallels that of the airways and another which courses within the connective tissue septa separating the segments.

Now that we have covered the anterior and posterior surface projections of the lungs and their lobes, we can proceed to examine the anterior and posterior surface projections of the pleural spaces around the lungs.

The pleural space that surrounds each lung is lined by a serous membrane (that is, a fluid-secreting membrane). Fig. 9-9 is a drawing of a coronal section (that is, a left-to-right section) of the right lung, the pleural space around it, and the surrounding parts of the rib cage, diaphragm, and mediastinal viscera when a person is breathing quietly. The serous membrane of the pleural space that lines the external surfaces of the lung (including the surfaces within the fissures) is called visceral pleura (the thin black line surrounding the lung represents visceral pleura). The serous membrane of the pleural space that lines the rib cage, diaphragm, and mediastinal viscera is called parietal pleura (the continuous, thin black line on the inner surface of the rib cage, the upper surface of the diaphragm, and the right surface of the mediastinal viscera represents parietal pleura).

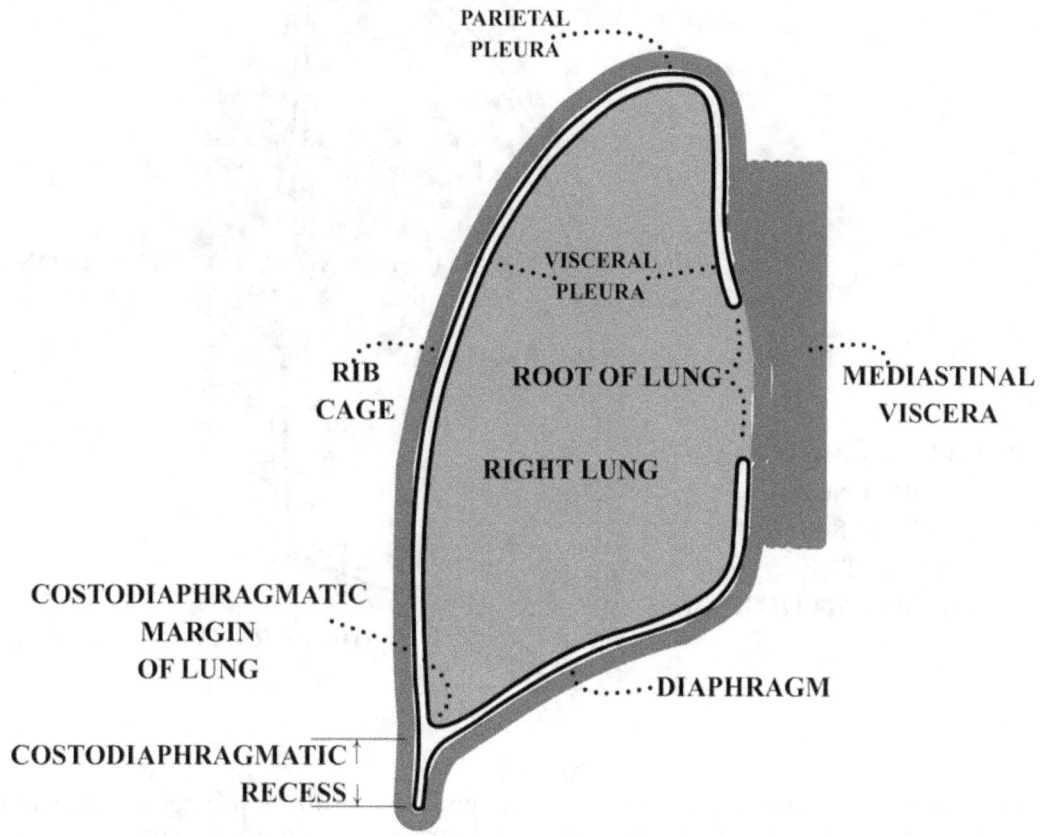

Fig. 9-9

Except for one area called the diaphragmatic recess, every area of the parietal pleura surrounding the lung directly faces visceral pleura. However, in the costodiaphragmatic recess, matching areas of parietal pleura face each other. The costodiaphragmatic recess area is named for the fact that it is a recess of the pleural space that extends around the outer margin of the base of the lung where the costal surface of the parietal pleura meets the diaphragmatic surface of the parietal pleura. The outer margin of the base of the lung at the top of the recess is called the costodiaphragmatic margin of the lung.

As already noted, Fig. 9-9 applies if a person is breathing quietly. When a person begins to breathe more deeply or vigorously, the lungs expand more with each breath, and the costodiaphragmatic margins of the lungs extend down into their costodiaphragmatic recesses. If one breathes in as deeply as possible, each lung expands to its greatest volume, and the costodiaphragmatic margin of the lung extends all the way down to the bottom of the costodiaphragmatic recess, as shown in Fig. 9-10.

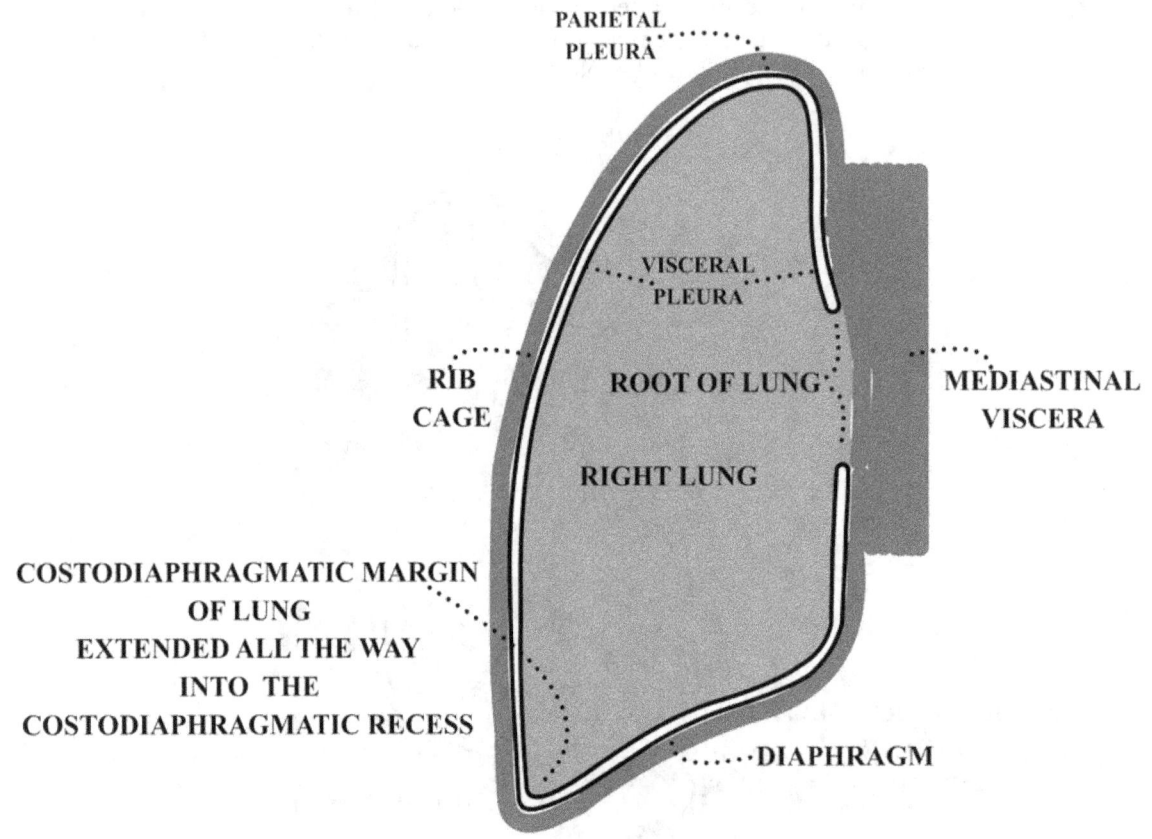

Fig. 9-10

40. The depth of the costodiaphragmatic recess remains constant as it extends all around the costodiaphragmatic margin of each lung. At every point along the costodiaphragmatic margin of both lungs, the bottom of the costodiaphragmatic recess lies 2 rib levels below that of the costodiaphragmatic margin of the lung (given that the patient is breathing quietly). Based upon your answers to question 35, complete the following statement:
When a person is breathing slowly,
   the midclavicular line intersects the bottom of the costodiaphragmatic recess
      at the level of the __8th__ rib,
   the midaxillary line intersects the bottom of the costodiaphragmatic recess
      at the level of the __10th__ rib, and
   the lateral border of the spine intersects the bottom of the costodiaphragmatic recess
      at the level of the __12th__ rib.

Knowledge of the surface projections of each lung's apex and fissures and the rib levels which overlie its costodiaphragmatic margin and the bottom of its costodiaphragmatic recess must always be recalled when physically examining or performing a procedure on a patient's chest. The following 4 questions illustrate the importance of this knowledge:

41. Where do you place the stethoscope when you want to listen to the breath sounds in the apex of a patient's lung?
_____ Immediately above the lateral third of the clavicle
__x__ Immediately above the medial third of the clavicle
_____ Over the anterior end of the 2nd intercostal space
_____ Over the intersection of the 2nd intercostal space with the midclavicular line

42. Cardiac failure (that is, the inability of the heart to maintain normal cardiac output) can lead to pleural effusion (accumulation of fluid within the pleural space). Pleural effusion that develops as a consequence of cardiac failure can sometimes be detected through auscultation or percussion of the lungs. With a patient suffering from cardiac failure who is sitting upright as you examine him/her, where would you expect almost all of the excess pleural fluid to be initially pooled in the pleural spaces of the lungs? Explain the anatomical basis of your answer.

You would expect almost all of the pleural effusion to be initially pooled in the lowest back part of the pleural spaces of the lungs. Because pleural fluid is free to move about within each lung's pleural space, gravity pools pleural effusion initially into the lowest region of the pleural space, which, when a person is seated upright, is the region of the pleural space that underlies the posterior rim of each lung's costodiaphragmatic margin. The bottom of the costodiaphragmatic recess of each pleural space is at its lowest level (the level of the 12th rib) in the back of the chest. In an adult, pleural effusion does not begin to reach the bottom of the costodiaphragmatic recess at the midaxillary line (which intersects the rib cage at the level of the 10th rib) until about 300-400 ml excess pleural fluid has accumulated in the pleural space.

43. When you place the stethoscope on a patient's right 4th intercostal space at a spot that intersects the right midclavicular line, you hear breath sounds in which lobe of the right lung? _____ upper lobe __x__ middle lobe _____ lower lobe

44. To perform a liver biopsy, a patient is first asked to lie in a supine position, that is, to lie on his or her back. The patient is asked to exhale as much as possible and then hold their breath as a needle is inserted into the chest at the 9th intercostal space along the midaxillary line. Explain why the needle will pass through the right lung's pleural space but not pierce the right lung's lower lobe.

When a person breathes quietly, the lower margin of the lung intersects the 8th rib at the midaxillary line. At the midaxillary line, the bottom of the costodiaphragmatic recess lies at the level of the 10th rib. When the patient breathes out as much as possible, the lungs occupy their least normal volume, and the lower margin of each lung rises somewhat above the level of the 8th rib along the midaxillary line. Consequently, insertion of the needle into the 9th intercostal space along the midaxillary line should place the needle at a level above the bottom of the right lung's costodiaphragmatic recess but below the lung's costodiaphragmatic margin.

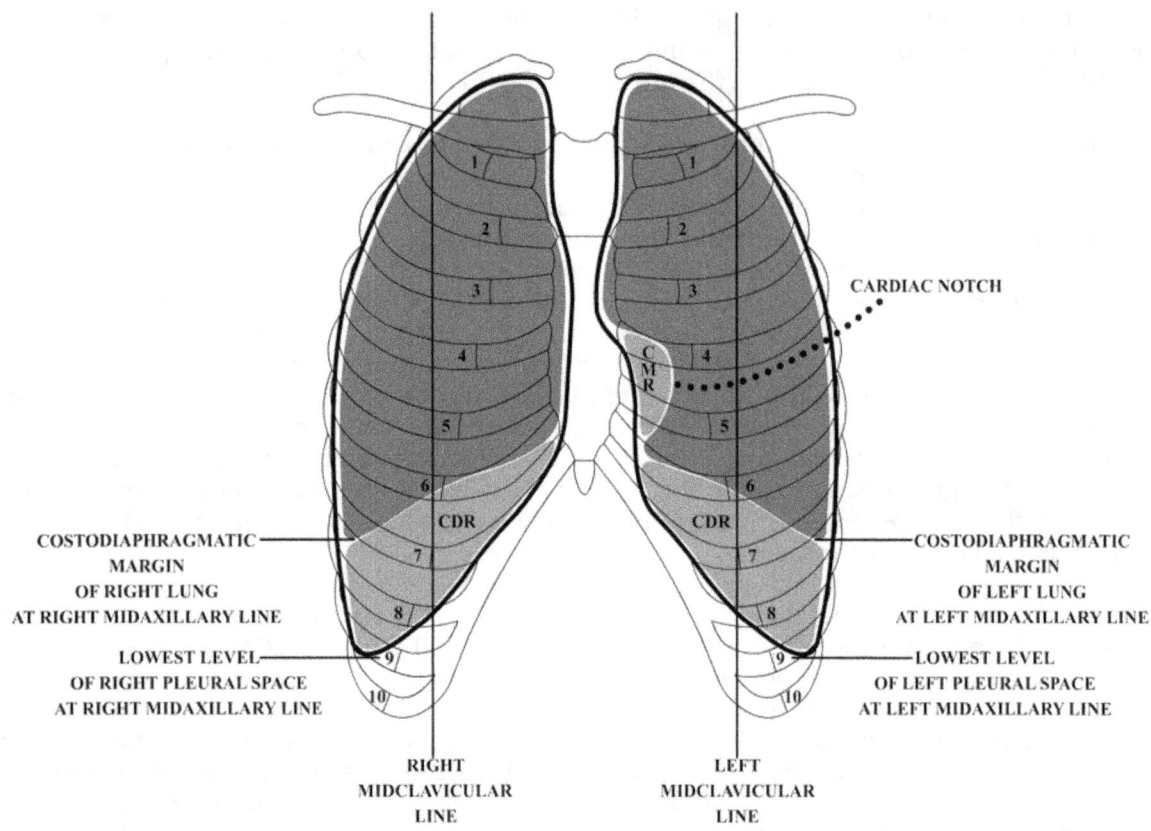

Fig. 9-11

Fig. 9-11 illustrates much of the chest anatomy that we have covered in the preceding 9 pages. Fig. 9-11 shows the midclavicular lines and the surface projections of the lungs and their pleural spaces onto the anterior chest wall. The dark gray areas represent the surface projections of the lungs onto the anterior chest wall; the light gray areas represent the surface projections of the costodiaphragmatic recesses onto the anterior chest wall. Observe that the lower margin of each lung intersects the midclavicular line at the level of the 6th rib and the midaxillary line at the level of the 8th rib. Notice also that the bottom of the costodiaphragmatic recess (CDR) for each lung intersects the midclavicular line at the level of the 8th rib and the midaxillary line at the level of the 10th rib. Finally, observe the cardiac notch of the medial border of the left lung that extends from the level of the left 4th costal cartilage to that of the left 6th costal cartilage.

Notice, however, that, during quiet respiration, there is a small pleural space recess called the costomediastinal recess (CMR) recess that borders the cardiac notch of the left lung. The margin of the left lung that forms the cardiac notch extends completely into the costomediastinal recess during forced respiration.

148

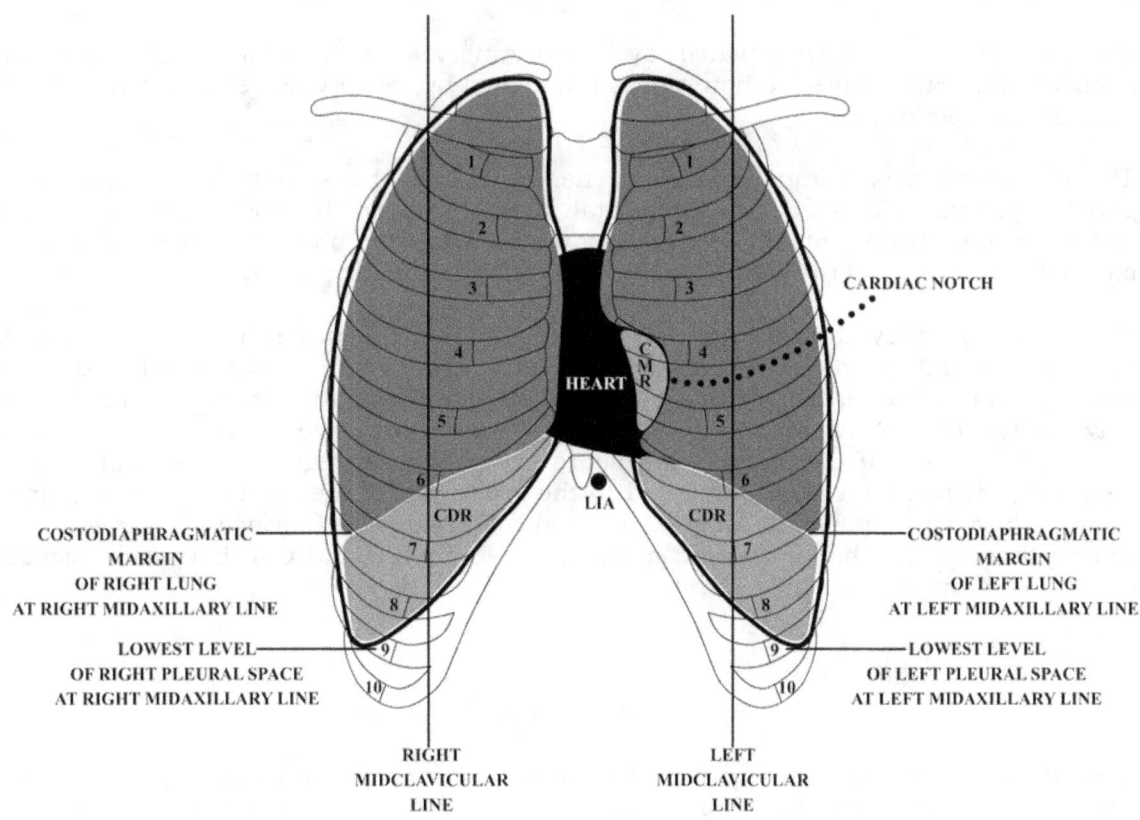

Fig. 9-12

Fig. 9-12 is identical to Fig. 9-11 except that the position of the heart is labelled in black in Fig. 9-12. In the chest, the heart is enveloped by a tissue sac called the pericardium. The pericardium encloses a space around the heart called the pericardial cavity. The pericardial cavity is lined by a fluid-secreting membrane called the serous pericardium. Under normal physiological conditions, the pericardial cavity in an adult contains 20 to 30 ml pericardial fluid. The pericardial cavity provides a lubricated free space that separates the heart from other mediastinal viscera and the lungs.

Accumulation of excess fluid in the pericardial cavity can occur in response to infection or the escape of blood from the heart following injury to the heart. A life-threatening condition called cardiac tamponade results if the fluid accumulates to the extent that it compresses and restricts the heart, thereby significantly reducing cardiac output. Fig. 9-12 shows that the most direct approach to inserting a needle into the pericardium to aspirate excess pericardial fluid is to point a needle toward the left shoulder as it is inserted into the skin overlying the left infrasternal angle (this site is marked by the black circle labelled LIA), which is the inferior angle between the lower margin of the left 7th costal cartilage and the left margin of the xiphoid process. This procedure is called pericardiocentesis. The needle is pointed toward the left shoulder as opposed to the right shoulder because the medial margins of both the left lung and its surrounding pleural space are laterally deviated from the level of the left 4th costal cartilage down to the level of the left 6th costal cartilage. The left side-oriented approach thus minimizes the risk that the needle will pass through the pleural space around the left lung or pierce the lingual region of the left lung.

149

45. Describe the sensory innervation of the visceral and parietal pleura.

Visceral pleura is not innervated by sensory nerves sensitive to touch, pain, and temperature. Parietal pleura, however, is innervated by sensory nerves sensitive to touch, pain, and temperature.

The intercostal nerves and the phrenic nerves supply the sensory innervation of the parietal pleura. Whereas the intercostal nerves supply the costal pleura and the circumferential region of the diaphragmatic pleura, the phrenic nerves supply the mediastinal pleura and the central region of the diaphragmatic pleura.

The pain evoked by the sensory nerves of the parietal pleura is generally sharp and superficial, and aggravated when air is breathed in. The pain may also be referred. If the costal pleura or the circumferential region of the diaphragmatic pleura is irritated, pain may be referred to the chest and abdominal walls along the course of the irritated intercostal nerves. If the mediastinal pleura or the central region of the diaphragmatic pleura is irritated, pain may be referred to the lower part of the neck or over the point of the shoulder [the cutaneous innervation of these regions is provided by nerves whose sensory fibers enter the spinal cord at the same levels (C3 and C4) that receive sensory nerve fibers from the phrenic nerves].

Mediastinum

The mediastinum is the median region of the chest. It is bordered anteriorly by the sternum, posteriorly by the thoracic vertebrae, and, on each side, by mediastinal pleura. The mediastinum is bordered above by the thoracic inlet and below by the diaphragm.

The superior mediastinum and the inferior mediastinum are the parts of the mediastinum, respectively, above and below the level of the sternal angle.

46. The sternal angle lies at the level of the _____ thoracic intervertebral disc.
   _____ 1st   _____ 2nd   _____ 3rd   _x_ 4th   _____ 5th   _____ 6th   _____ 7th

The inferior mediastinum is divisible into three regions: The anterior mediastinum (which is the region of the inferior mediastinum between the sternum and the pericardium), the middle mediastinum (which is the mediastinal region consisting of the pericardium and its contents and the tracheal bifurcation), and the posterior mediastinum (which is the region of the inferior mediastinum between the pericardium and the spine).

Indicate the regions of the mediastinum either occupied or traversed by the following viscera:

Viscus	Superior Mediastinum	Anterior Mediastinum	Middle Mediastinum	Posterior Mediastinum
47. Arch of aorta	x			
48. Ascending aorta			x	
49. Azygos vein			x	x
50. Brachiocephalic trunk	x			
51. Brachiocephalic veins	x			
52. Carina			x	
53. Descending thoracic aorta				x
54. Esophagus	x			x
55. Heart			x	
56. Internal thoracic arteries	x			
57. Left common carotid artery	x			
58. Left recurrent laryngeal nerve	x			
59. Left subclavian artery	x			
60. Main stem bronchi			x	
61. Phrenic nerves	x		x	
62. Pulmonary arteries			x	
63. Pulmonary trunk			x	
64. Pulmonary veins			x	
65. Superior vena cava	x		x	
66. Thoracic duct	x			x
67. Thymus	x	x		
68. Trachea	x			
69. Vagus nerves	x			x

ABDOMEN – Part A: Questions

Dissection of the abdomen in gross lab focuses on identification of
  (1) the muscles, rectus sheath, arteries, and nerves of the anterior abdominal wall,
  (2) the parts of the abdominal organs and glands, their arterial supply, and venous drainage,
  (3) the ureters,
  (4) the gonadal arteries and veins, and
  (5) the muscles and nerves of the posterior abdominal wall.

The abdominal anatomy most frequently applied in clinical practice is the surface anatomy relevant to the physical examination of the abdominal organs and the inguinal region of the anterior abdominal wall.

Anterior Abdominal Wall

1. Describe the palpable landmarks of the upper and lower borders of the anterior abdominal wall.

2. On each side of the body, the inguinal ligament marks the border between the abdomen and the front of the thigh. What are the inguinal ligaments and to what are they attached?

3. Identify the dermatome of each of the following anterior abdominal wall skin regions:
   _____ Skin overlying the xiphoid process
   _____ Skin surrounding the umbilicus
   _____ Skin overlying the inguinal region of the anterior abdominal wall

The anterior abdominal wall region that immediately borders the anterior thigh is called the inguinal region. The inguinal region has a passageway in it through which the spermatic cord in the male and the round ligament of the uterus extend between the abdomen and the external genitalia. Structurally, the inguinal canal is, in effect, a defect in the anterior abdominal wall that renders the inguinal region especially susceptible to herniation.

4. The formation of the inguinal canal spans a time period in the fetal life of both sexes extending from the 5th to the 32nd week. The development which occurs up to the 28th week is very similar in both sexes. Describe the events common to both sexes by which the inguinal canal is formed from the 5th to the 28th week.

5. Describe the events in the male fetus associated with the descent of the testis through the inguinal canal into the scrotum from the 28th week of fetal development to about the 4th postnatal week.

A hernia is most commonly defined as a protrusion of a tissue or organ through a wall by which it is normally contained. However, in order to avoid the semantic argument of whether the term hernia applies to the protrusion versus the wall opening itself, it is best to recognize that the common denominator to all hernia definitions is the anatomical defect in the supporting structures. Bearing this in mind, a hernia of an abdominal wall is probably best defined as a weakness or opening in the muscular and fascial layers through which a contained tissue or organ may protrude.

An abdominal wall hernia almost always consists of three parts, which, proceeding from the innermost to the most superficial part, are the contents of the hernia, its sac, and the sac's coverings. The contents of a hernia may include any tissue or organ in the abdominal or pelvic cavity, commonly a segment of the small or large intestine. The sac of a hernia consists of an outpouching of parietal peritoneum; this peritoneum outpouching envelops the contents of the hernia as they protrude through the abdominal wall region (parietal peritoneum is any region of peritoneum that lines the inner surface of an abdominal wall region). The neck of the sac of a hernia is the tapered, proximal end of the sac (it is the end at which the sac is continuous with the parietal peritoneum). The sac coverings of a hernia are the tissue layers of the abdominal wall region (superficial to the peritoneum) through which the contents of the hernia and its sac protrude.

6. There are two types of inguinal hernias: indirect inguinal hernias and direct inguinal hernias. Describe the embryological derivation of indirect inguinal hernias and the location of the neck of the sac of all indirect inguinal hernias relative to the deep inguinal ring and the inferior epigastric artery.

7. Describe the derivation of direct inguinal hernias and the location of the neck of the sac of all direct inguinal hernias relative to the inferior epigastric artery.

8. What are the structures that form the roof and floor of the inguinal canal?

9. What is a femoral hernia? Describe the derivation of femoral hernias.

10. What is a caput medusae?

Abdominopelvic and Peritoneal Cavities

11. Describe the nature of the abdominopelvic and peritoneal cavities and their relationship to each other in the lower part of the trunk of the body.

12. What is the distinction between parietal peritoneum and visceral peritoneum?

13. What is the distinction between retroperitoneal and intraperitoneal organs?

14. What are peritoneal ligaments?

15. What are the greater and lesser sacs of the peritoneal cavity, and where do they communicate with each other?

16. What are the borders of the epiploic foramen?

Regions of the Abdominal Cavity

When recording the results of a patient's physical examination, it is sometimes important to record the location of abdominal pain or an abdominal mass. Physicians use one of two methods to divide the abdomen into well-defined sub-regions. Both methods employ the mental projection of vertical and horizontal planes onto and through the anterior abdominal wall of the patient.

17. One of the methods divides the abdomen into four, roughly equal regions called quadrants. How is the abdominal cavity divided into quadrants?

18. The other method divides the abdomen into nine, roughly equal regions. How is the abdominal cavity divided into the 9 roughly equal regions?

19. A person unfamiliar with the surface anatomy of the chest and abdomen might be surprised that each method for subdividing the abdomen places the highest abdominal regions deep to the lower parts of the rib cage. The diaphragm marks the highest level of the abdomen in the body. As a check on what you learned when you answered question 14 in the chapter on the chest, can you recall which of the four drawings in Fig. 10-16 most accurately represents the anterior surface projection of the diaphragm in a healthy adult lying supine on an examination table and breathing quietly?

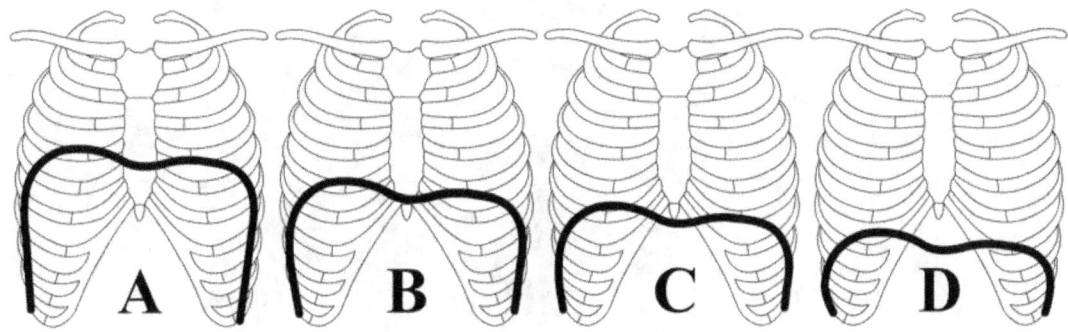

Fig. 10-16

20. The abdominal aorta and inferior vena cava are the largest abdominal blood vessels. Fig. 10-17 shows that the abdominal aorta begins at the level of the body of the 12th thoracic vertebra (where it passes through the aortic opening of the diaphragm) and then descends within the abdomen slightly to the left of the midline. Fig. 10-17 also shows that the inferior vena cava ascends within the abdomen slightly to the right of the midline and ends at the level of the body of the 8th thoracic vertebra (where it passes through the caval opening of the diaphragm). Indicate the vertebral level at which the abdominal aorta ends (upon bifurcating into the paired common iliac arteries) and the vertebral level at which the inferior vena cava begins (from the union of the paired common iliac veins).

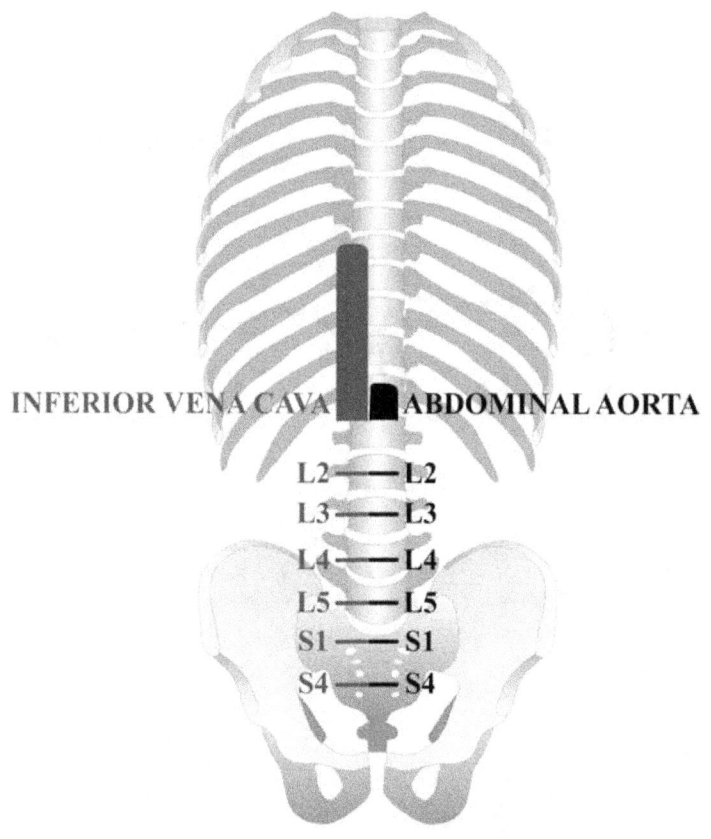

INFERIOR VENA CAVA   ABDOMINAL AORTA

L2———L2
L3———L3
L4———L4
L5———L5
S1———S1
S4———S4

Fig. 10-17

21. The kidneys are the largest retroperitoneal organs. They lie nestled in the paravertebral gutters, which are the gutters alongside the spine. The right kidney lies at a level slightly lower than that of the left kidney because of the prominence of the liver on the right side of the body. Which of the 6 drawings in Fig. 10-18 most accurately represents the positions of the kidneys in a healthy adult who is lying in a supine position on an examination table and breathing quietly?

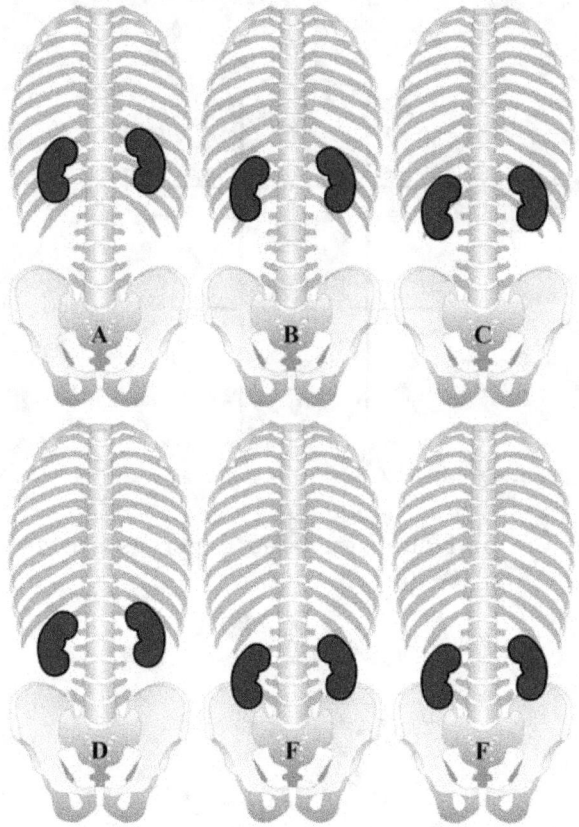

Fig. 10-18

22. The adrenal glands are retroperitoneal viscera that have a constant spatial relationship relative to the kidneys. Describe this spatial relationship.

23. The largest secondarily retroperitoneal viscera of the abdomen are the ascending colon and descending colon. Outline their locations in Fig. 10-19.

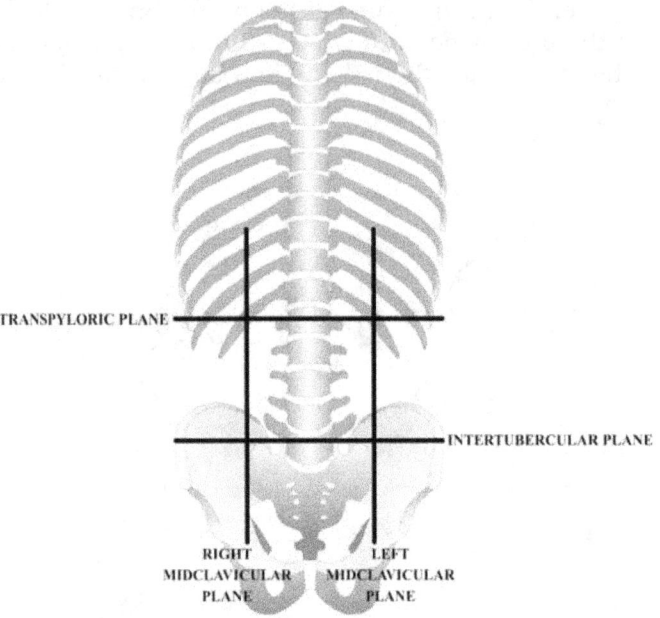

Fig. 10-19

24. Almost all the duodenum and almost all the pancreas are secondarily retroperitoneal. Outline their locations in Fig. 10-20.

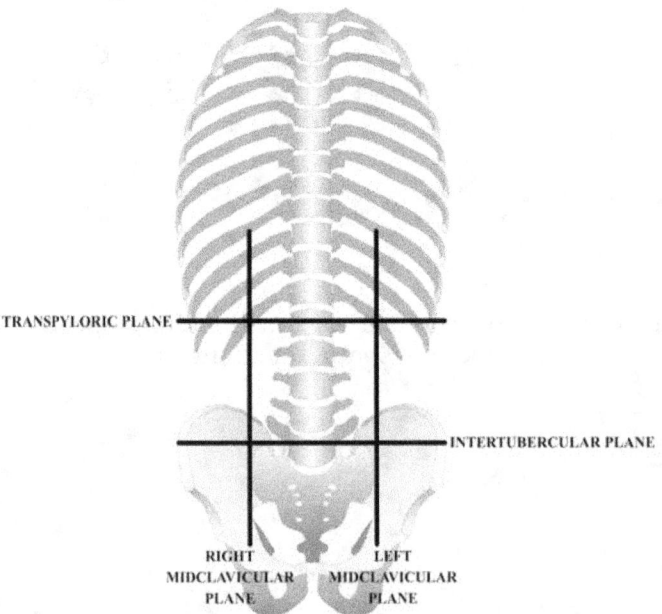

Fig. 10-20

158

25. The liver and gallbladder have a constant relationship relative to each other and relatively constant relationships to the regions of the abdomen. Explain the anatomical basis of these relationships, and outline the location of the liver in Fig. 10-21.

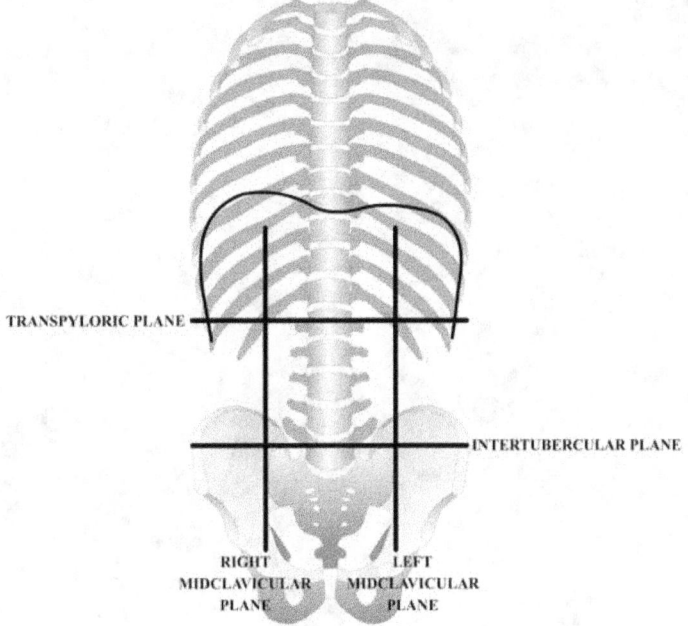

Fig. 10-21

26. Which part of the liver is palpable during a physical examination of a patient?

27. Describe how percussion can be used to assess the size of the liver during a physical examination of a patient.

28. Part of the right lobe of the liver can be anatomically subdivided into lobes called the quadrate and caudate lobes of the liver. Which surface features of the liver are used to define the borders of the quadrate and caudate lobes of the liver?

29. What is the distinction between the anatomical division of the liver into right and left lobes by the falciform ligament and the functional division of the liver into right and left sides on the basis of its blood supply and bile production?

30. What is Murphy's sign?

31. The spleen is an intraperitoneal organ that has a relatively constant relationship to the rib cage. Explain the anatomical basis of this relationship, and identify which of the drawings in Fig. 10-23 most accurately represents the position of a normal sized spleen relative to the posterior rib cage.

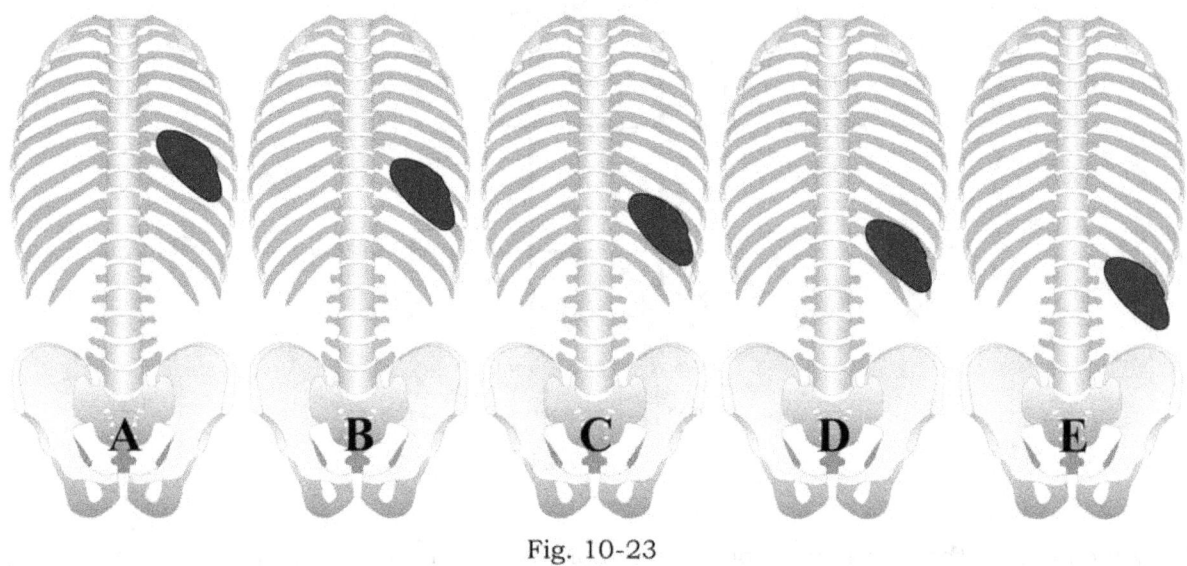

Fig. 10-23

32. Is a normal sized spleen palpable during a physical examination of an adult patient?

33. The stomach is an intraperitoneal organ whose openings with the lower end of the esophagus (the cardiac opening) and the beginning of the duodenum (the pyloric opening) have constant relationships with the spine. Describe these relationships and outline the general position of an empty stomach in an adult in Fig. 10-24.

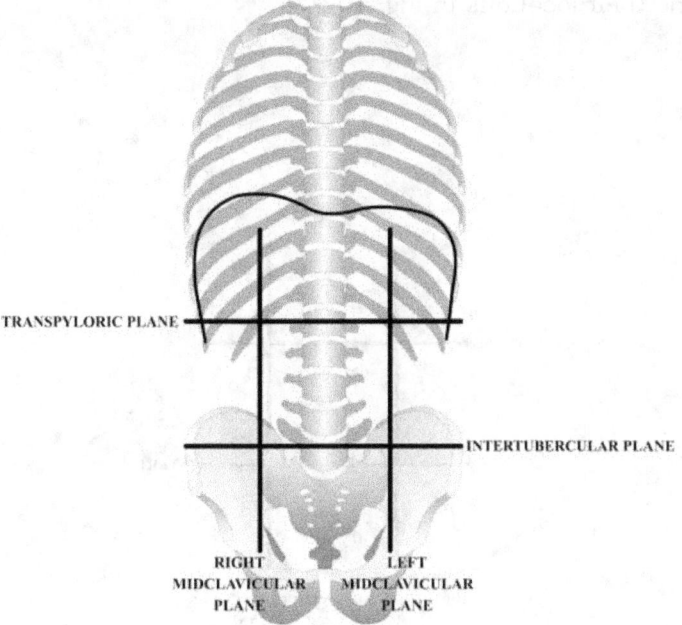

Fig. 10-24

34. The cecum is an intraperitoneal segment of the large intestine. Outline the location of the cecum in Fig. 10-25.

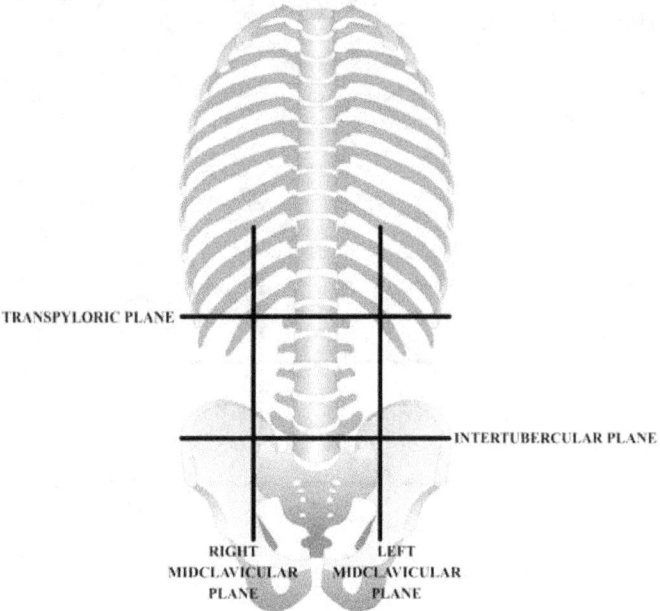

Fig. 10-25

161

35. The appendix is an intraperitoneal organ. Explain the anatomical basis of its variable position in the abdomen.

36. The transverse colon and the sigmoid colon are both intraperitoneal segments of the large intestine. Outline their locations in Fig. 10-27.

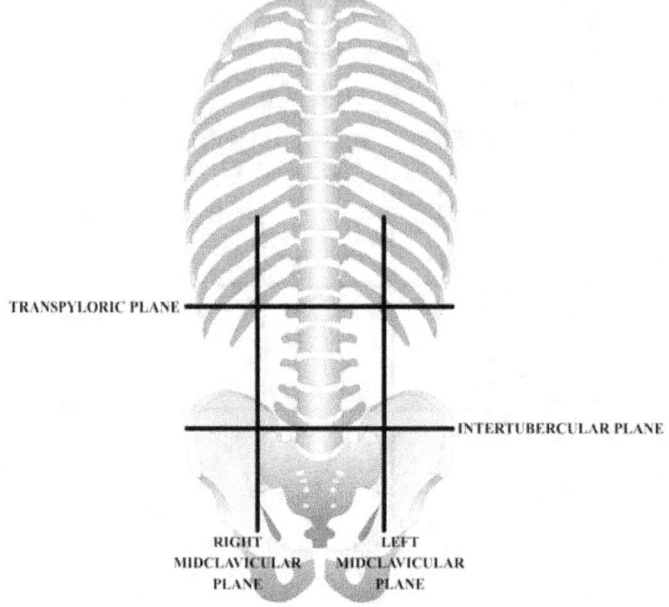

Fig. 10-27

37. The jejunum and ileum are both intraperitoneal segments of the small intestine. Outline their general location in Fig. 10-28.

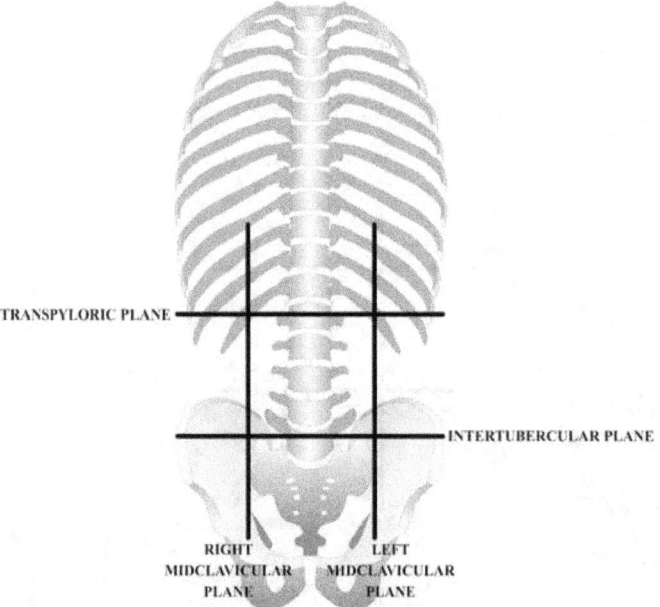

Fig. 10-28

Types of Pain Associated with Disease and Injury of Abdominal Viscera

38. Disease or injury of an abdominal organ or gland can produce three types of pain: visceral pain, referred pain, and/or somatic pain. What is visceral pain?

39. Which abdominal viscera, when diseased or injured, typically produce visceral pain in (a) the epigastric region, (b) the umbilical region, and (c) the hypogastric region?

40. What is referred pain?

41. What is somatic pain?

42. What is rebound tenderness?

END OF QUESTIONS IN PART A OF THE CHAPTER ON THE ABDOMEN

164

ABDOMEN – Part B: Questions and Answers

Dissection of the abdomen in gross lab focuses on identification of
- (1) the muscles, rectus sheath, arteries, and nerves of the anterior abdominal wall,
- (2) the parts of the abdominal organs and glands, their arterial supply, and venous drainage,
- (3) the ureters,
- (4) the gonadal arteries and veins, and
- (5) the muscles and nerves of the posterior abdominal wall.

The abdominal anatomy most frequently applied in clinical practice is the surface anatomy relevant to the physical examination of the abdominal organs and the inguinal region of the anterior abdominal wall.

Anterior Abdominal Wall

2. Describe the palpable landmarks of the upper and lower borders of the anterior abdominal wall.

    The xiphoid process and the costal margins of the rib cage mark the upper border of the anterior abdominal wall.

    The iliac crests, anterior superior iliac spines, pubic tubercles, and pubic crests of the coxal bones together with the pubic symphysis mark the lower border of the anterior abdominal wall.

2. On each side of the body, the inguinal ligament marks the border between the abdomen and the front of the thigh. What are the inguinal ligaments and to what are they attached?

    Three muscles (external oblique, internal oblique, and transversus abdominis) form most of the flank, or side, region of the anterior abdominal wall on each side of the body. Each of these flank muscles has a thin, sheet-like insertion tendon; thin, sheet-like tendons are called aponeurotic tendons. External oblique is the most superficial flank muscle; its aponeurotic tendon has a lower free margin which is attached laterally to the anterior superior iliac spine and medially to the pubic tubercle of the coxal bone. This lower free margin of external oblique's aponeurotic tendon is called the inguinal ligament because it is attached at its end to bony landmarks.

3. Identify the dermatome of each of the following anterior abdominal wall skin regions:
    __T7__ Skin overlying the xiphoid process
    _T10__ Skin surrounding the umbilicus
    __L1__ Skin overlying the inguinal region of the anterior abdominal wall

The anterior abdominal wall region that immediately borders the anterior thigh is called the inguinal region. The inguinal region has a passageway in it through which the spermatic cord in the male and the round ligament of the uterus extend between the abdomen and the external genitalia. Structurally, the inguinal canal is, in effect, a defect in the anterior abdominal wall that renders the inguinal region especially susceptible to herniation.

4. The formation of the inguinal canal spans a time period in the fetal life of both sexes extending from the 5th to the 32nd week. The development which occurs up to the 28th week is very similar in both sexes. Describe the events common to both sexes by which the inguinal canal is formed from the 5th to the 28th week.

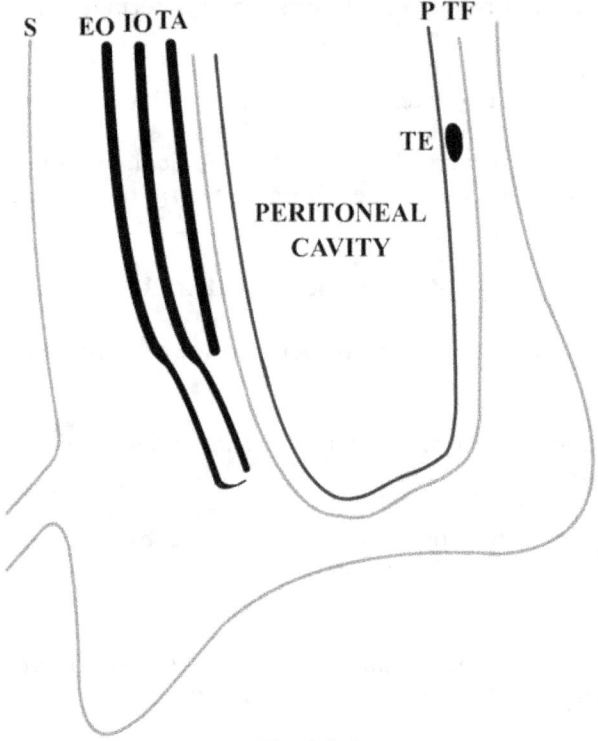

Fig. 10-1

Fig. 10-1 shows a coronal section (that is, a section that extends from front to back) of the lower region of the abdomen in a male fetus during the 5th week of gestation. The relative sizes and spaces among structures are distorted for the purpose of clarifying the events by which the inguinal canal is formed. The drawing shows major tissue layers of the flank region of the anterior abdominal wall: skin (S), the external oblique muscle (EO), the internal oblique muscle (IO), the transversus abdominis muscle (TA), the transversalis fascia (TF), and finally, the deepest tissue layer, the peritoneum (P). The peritoneum is the serous (that is, fluid-secreting) membrane that lines the entire peritonal cavity. The formation of the gonads (the testes in the male and the ovaries in the female) begins in both sexes during the 5th week of fetal development. Fig. 10-1 shows that, in the male, the testes (TE) arise in the lower region of the posterior abdominal wall, in a layer of connective tissue sandwiched between the peritoneum and the transversalis fascia.

Fig. 10-2, which focuses on the abdominal region where the inguinal canal will be formed, shows the event which initiates the formation of the inguinal canal: the forward protrusion of a peritoneal evagination called the vaginal process (VP).

Fig. 10-3 shows the next event in the formation of the inguinal canal: the protrusion of the vaginal process into the overlying transversalis fascia. As the vaginal process pushes forward into the transversalis fascia, it acquires a tubular fascial covering called the internal spermatic fascia (ISP) (the name is the same for both sexes). The rounded margin where the internal spermatic fascia is continuous with the transversalis fascia is called the deep inguinal ring. The deep inguinal ring is the lateral, or deep, end of the inguinal canal.

166

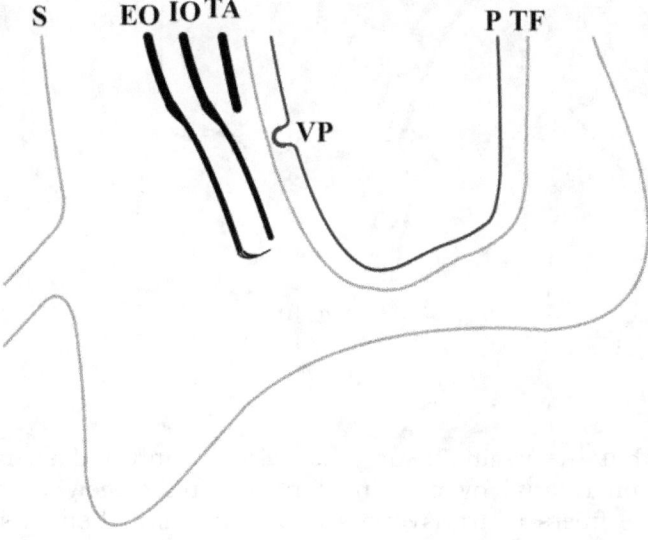

Fig. 10-2

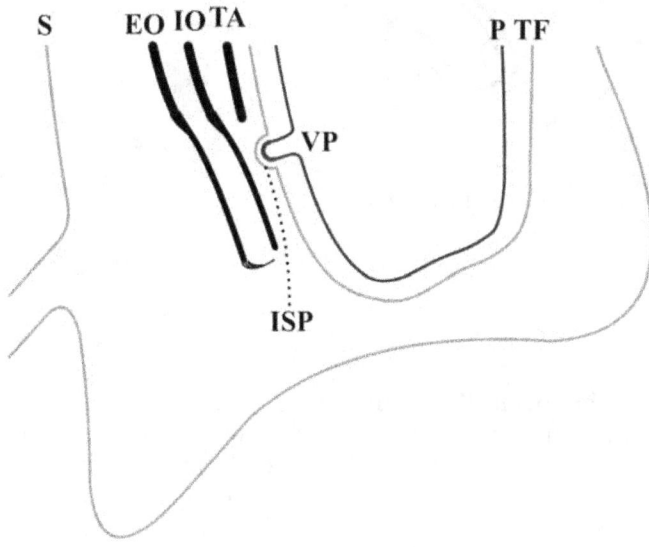

Fig. 10-3

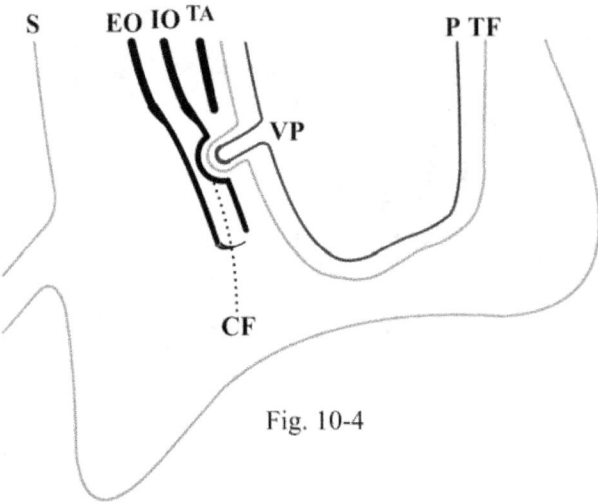

Fig. 10-4

Fig. 10-4 shows that the vaginalis process, with its internal spermatic fascia covering, generates the inguinal canal by pushing forward and somewhat medially, first passing beneath the muscle fibers of transversus abdominis and then pushing forward into the internal oblique. As the vaginal process pushes forward into internal oblique, it acquires a second tubular fascial covering called the cremasteric fascia (CF).

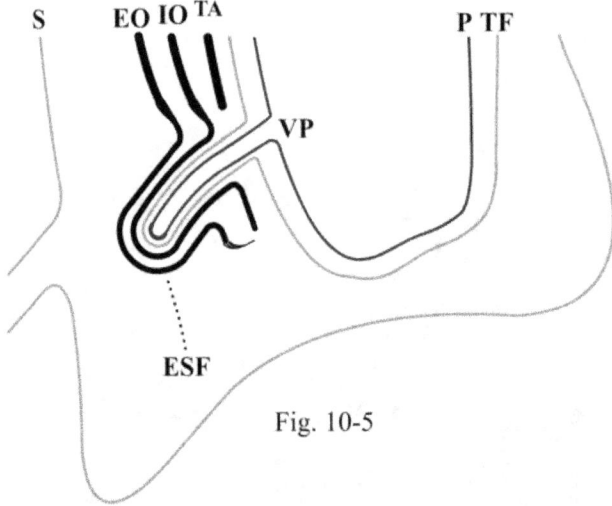

Fig. 10-5

Fig. 10-5 shows that the formation of the inguinal canal ends as the vaginal process, with its internal spermatic and cremasteric fascia coverings, pushes forward into external oblique's tendon to acquire a third tubular fascial covering called the external spermatic fascia (ESF) (the name is the same for both sexes). The triangular margin where the external spermatic fascia is continuous with the external oblique's tendon is called the superficial inguinal ring. The superficial inguinal ring is the medial, or superficial, end of the inguinal canal, and it lies immediately above and medial to the pubic tubercle of the coxal bone.

5. Describe the events in the male fetus associated with the descent of the testis through the inguinal canal into the scrotum from the 28th week of fetal development to about the 4th postnatal week.

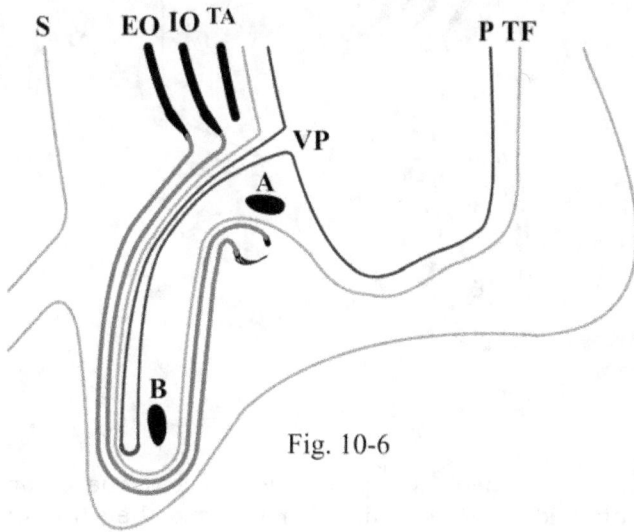

Fig. 10-6

Fig. 10-6 shows the protrusion of the vaginal process, with its internal spermatic, cremasteric, and external spermatic fascia coverings, into the developing scrotum of a male fetus during the 28th week of fetal development. By the 28th week of fetal development, the testis has usually migrated into the deep inguinal ring of the inguinal canal, which is marked position A in Fig. 10-6. It then descends through the inguinal canal along a pathway that is directly behind, but topologically superficial to, the vaginal process; this pathway is enclosed within the fascial coverings that the vaginal process acquired as it formed the inguinal canal. Accordingly, by the time that the testis passes through the superficial inguinal ring and into the developing scrotum (an event which commonly occurs during the 32nd week of fetal age), it too is ensheathed within the same fascial coverings that ensheath the vaginal process. The testis occupies position B upon completing its descent into the scrotum.

As the testis descends all the way from the site of its genesis in the lower posterior abdominal wall through the inguinal canal and into the scrotum, it extends along with itself the nerves and blood and lymphatic vessels that initially served it when it was forming in the posterior abdominal wall. It also extends along with itself the vas deferens, the tubelike vessel which, during sexual climax, transmits sperm from the epididymis to the ejaculatory duct in the prostate gland. This collection of structures (the vas deferens, nerves, and blood and lymphatic vessels) that extends from the testis is referred to as the spermatic cord.

Once the testis has descended into the scrotum, the formation of the inguinal canal and its contents in the male is virtually finished. The only developmental event of importance still to occur is the sealing-up of the distal end of the vaginal process (the end part in front of the testis) and the degeneration of the remaining, proximal part during the first few weeks following birth, as shown in Fig. 10-7. The only postnatal, gross remnant of the vaginal process is thus a closed sac called the tunica vaginalis (TV) which covers all but the posterior surface of the testis.

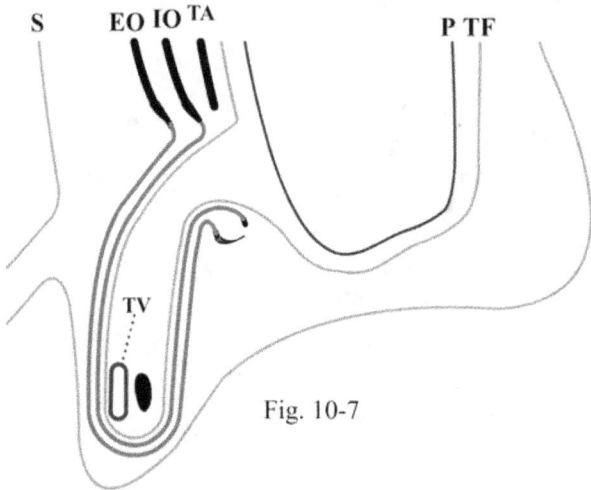

Fig. 10-7

A hernia is most commonly defined as a protrusion of a tissue or organ through a wall by which it is normally contained. However, in order to avoid the semantic argument of whether the term hernia applies to the protrusion versus the wall opening itself, it is best to recognize that the common denominator to all hernia definitions is the anatomical defect in the supporting structures. Bearing this in mind, a hernia of an abdominal wall is probably best defined as a weakness or opening in the muscular and fascial layers through which a contained tissue or organ may protrude.

An abdominal wall hernia almost always consists of three parts, which, proceeding from the innermost to the most superficial part, are the contents of the hernia, its sac, and the sac's coverings. The contents of a hernia may include any tissue or organ in the abdominal or pelvic cavity, commonly a segment of the small or large intestine. The sac of a hernia consists of an outpouching of parietal peritoneum; this peritoneum outpouching envelops the contents of the hernia as they protrude through the abdominal wall region (parietal peritoneum is any region of peritoneum that lines the inner surface of an abdominal wall region). The neck of the sac of a hernia is the tapered, proximal end of the sac (it is the end at which the sac is continuous with the parietal peritoneum). The sac coverings of a hernia are the tissue layers of the abdominal wall region (superficial to the peritoneum) through which the contents of the hernia and its sac protrude.

6.   There are two types of inguinal hernias: indirect inguinal hernias and direct inguinal hernias. Describe the embryological derivation of indirect inguinal hernias and the location of the neck of the sac of all indirect inguinal hernias relative to the deep inguinal ring and the inferior epigastric artery.

In some newborn infants (especially in those born prematurely), the proximal part of the vaginal process does not degenerate during the first few weeks following birth. The entire vaginal process persists and remains patent (that is, its interior remains continuous with the peritoneal cavity). This continuity provides a passageway by which a part of an abdominal organ (generally a loop of ileum or jejunum) can pass through the deep inguinal ring and extend into the inguinal canal; such an occurrence is an indirect inguinal hernia. The neck of the sac of an indirect inguinal hernia is thus always located at the deep inguinal ring. Indirect inguinal hernias have an embryological derivation because they almost always occur in individuals with a patent vaginal process.

170

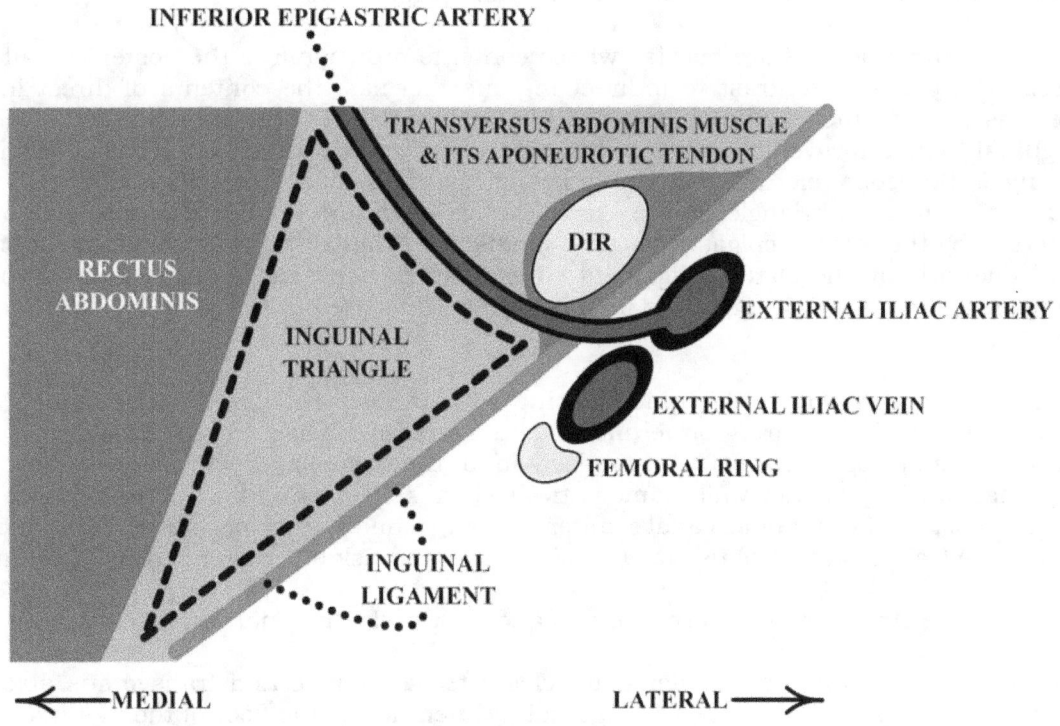

INFERIOR EPIGASTRIC ARTERY

TRANSVERSUS ABDOMINIS MUSCLE
& ITS APONEUROTIC TENDON

DIR

RECTUS
ABDOMINIS

INGUINAL
TRIANGLE

EXTERNAL ILIAC ARTERY

EXTERNAL ILIAC VEIN

FEMORAL RING

INGUINAL
LIGAMENT

←—MEDIAL   LATERAL—→

Fig. 10-8

Fig. 10-8 is a posterior view of the inguinal region of the anterior abdominal wall on the right side of the body. This is the view that surgeons have when they use laparoscopy to directly visualize the posterior surface of the inguinal region and repair inguinal hernias. The two deepest tissue layers of the inguinal region, namely, the peritoneum and the transversalis fascia, are not shown. Observe that the inguinal region is bordered inferiorly by the inguinal ligament and medially by the rectus abdominis muscle. Notice also that transversus abdominis (the deepest flank muscle) and its aponeurotic tendon together form a major structural feature of the inguinal region except for the area around the deep inguinal ring (DIR). Observe that immediately below the inguinal ligament lie the external iliac artery and the external iliac vein. As the external iliac artery extends inferiorly and beneath the inguinal ligament to enter the uppermost part of the thigh, it becomes continuous with the femoral artery. The segment of the external iliac vein shown in Fig. 10-8 is the continuation of the femoral vein as it extends superiorly and below the inguinal ligament to enter the abdomen. Finally, notice that the external iliac artery gives rise to a branch called the inferior epigastric artery that extends through the inguinal region as it ascends the anterior abdominal wall.

Fig. 10-8 shows that because the neck of the sac of an indirect inguinal hernia is always located at the deep inguinal ring, it is also lateral to the inferior epigastric artery. Surgeons use this latter relationship to distinguish between indirect and direct inguinal hernias.

The contents of indirect inguinal hernias may ultimately traverse the inguinal canal and extend into the scrotum of a male or the labium majus of a female. Because the contents of indirect inguinal hernias lie within the relatively narrow inguinal canal upon passing through the deep inguinal ring, they are always at risk of incarceration (entrapment within the inguinal canal) and strangulation (loss of blood supply).

171

7. Describe the derivation of direct inguinal hernias and the location of the neck of the sac of all direct inguinal hernias relative to the inferior epigastric artery.

Direct inguinal hernias are hernias whose contents protrude into the posterior wall of the inguinal canal. In contrast to indirect inguinal hernias, the contents of direct inguinal hernias do not enter, or lie, within the inguinal canal. The neck of the sac of a direct inguinal hernia always lies medial to the inferior epigastric artery and within the inguinal triangle. Surgeons refer to the inguinal triangle as Hesselbach's triangle. As shown in Fig. 10-8, the inguinal triangle is the triangular area within the inguinal region bordered laterally by the inferior epigastric artery, medially by the lateral edge of rectus abdominis, and inferiorly by the inguinal ligament. Because the contents of direct inguinal hernias bulge into the posterior wall of the inguinal canal, they are not at risk of incarceration and strangulation.

Conditions and activities which increase intra-abdominal pressure, such as obesity, heavy lifting, and straining during bowel movements increase tension within the inguinal region of the anterior abdominal wall, particularly in (a) the tissue layer which forms most of the inguinal canal's posterior wall, namely, transversalis fascia, and (b) the tissue layer which forms most of the inguinal canal's anterior wall, namely, external oblique's aponeurotic tendon. Weakness of these tissue layers increases the risk of a direct inguinal hernia.

8. What are the structures that form the roof and floor of the inguinal canal?

The lower free borders of the flank muscles internal oblique and transversus abdominis form the inguinal canal's roof. The inguinal ligament forms the floor of the inguinal canal.

9. What is a femoral hernia? Describe the derivation of femoral hernias.

Notice in Fig. 10-8 that a small opening called the femoral ring lies medial to the external iliac vein. The femoral ring is the upper end of a very narrow passageway called the femoral canal that extends between the abdomen and the upper border of the thigh. The femoral canal contains 1 or 2 lymph nodes called the deep inguinal lymph nodes. A femoral hernia is a hernia whose contents protrude through the femoral ring into the femoral canal. The neck of the sac of a femoral hernia always lies immediately lateral and inferior to the pubic tubercle, and extends inferior to the inguinal ligament.

As with direct inguinal hernias, conditions and activities which increase intra-abdominal pressure, such as obesity, heavy lifting, and straining during bowel movements increase the likelihood of femoral hernias. Because the contents of femoral hernias extend within the very narrow femoral canal upon passing through the femoral ring, they, like indirect inguinal hernias, are always at risk of incarceration and strangulation.

10. What is a caput medusae?

Almost all the veins of the anterior abdominal wall conduct blood toward either the superior vena cava or the inferior vena cava. However, there are several relatively small, superficial veins around the umbilicus called the para-umbilical veins which conduct blood toward the portal vein; the portal vein is a vein in the abdomen which conducts venous blood drained form the digestive tract directly into the liver. The para-umbilical veins extend toward the umbilicus from the perimeter of the region around the umbilicus. Under normal conditions, the presence of the para-umbilical veins in the superficial fascia is not apparent upon inspection of the anterior abdominal wall. However, under certain conditions (such as instances in which liver disease markedly elevates blood pressure within the portal vein), the para-umbilical veins become markedly distended and form a radiating pattern of varicose veins about the umbilicus. In remembrance of Medusa, a Greek mythological character whose head was covered with a multitude of snakes, the pattern of varicose, para-umbilical veins is termed a caput medusae (the head of Medusa).

Abdominopelvic and Peritoneal Cavities

11. Describe the nature of the abdominopelvic and peritoneal cavities and their relationship to each other in the lower part of the trunk of the body.

The abdominal and pelvic regions of the body share a common inner cavity called the abdominopelvic cavity (Fig. 10-9). The abdominopelvic cavity is enclosed by walls, which consist of the diaphragm superiorly, the anterior abdominal wall anteriorly, the posterior abdominal wall posteriorly, and the floor of the pelvis inferiorly. The boundary between the abdominal and pelvic cavities is a plane called the pelvic inlet.

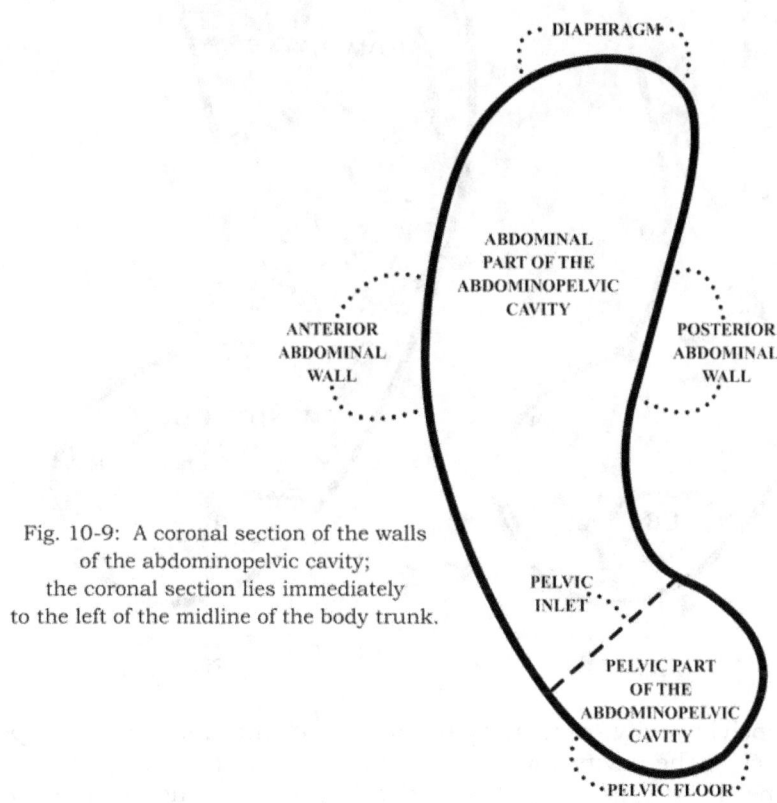

Fig. 10-9: A coronal section of the walls of the abdominopelvic cavity; the coronal section lies immediately to the left of the midline of the body trunk.

173

The abdominopelvic cavity houses all the abdominal and pelvic organs. Fig. 10-10 shows the approximate locations of several abdominal organs [the liver, stomach, pancreas (P), duodenum (D), a segment of the jejunum (J), a segment of the ileum (I), and a segment of the transverse colon (TC)] and two pelvic organs [the urinary bladder (UB) and the rectum (R)] in the coronal section of the abdominopelvic cavity shown in Fig. 10-9.

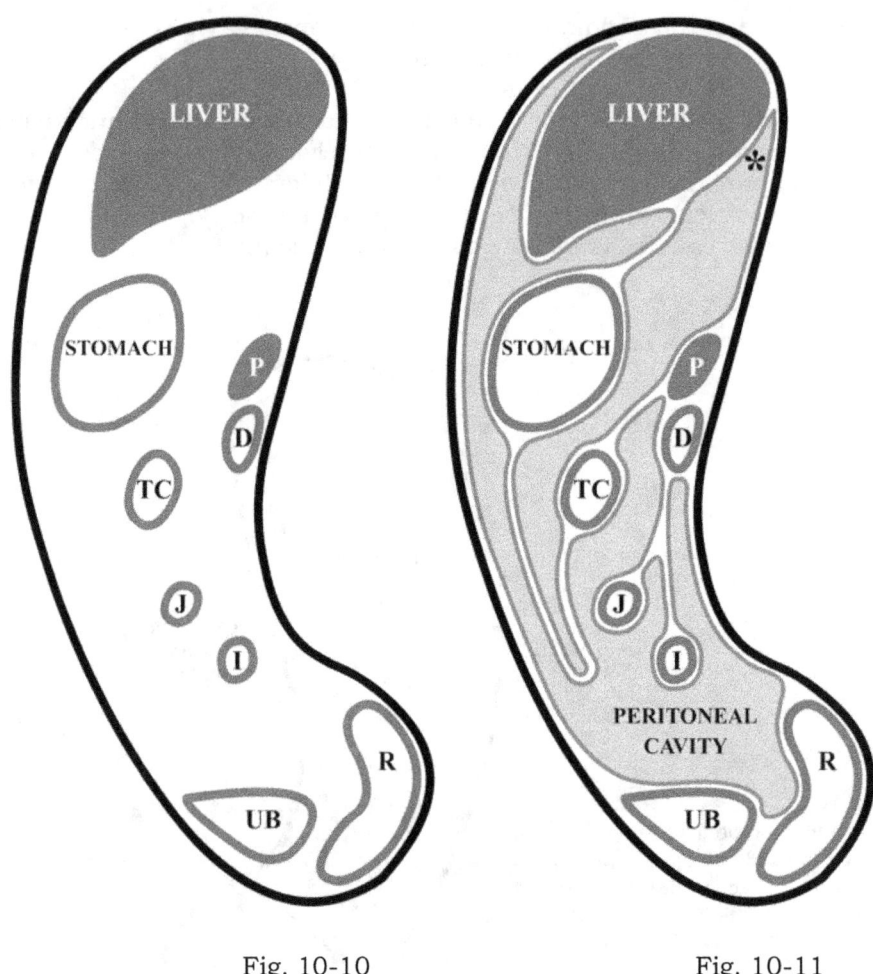

Fig. 10-10   Fig. 10-11

Within the abdominopelvic cavity, a continuous, serous membrane called the peritoneum lines the inner surfaces of the cavity's walls as well as parts of the external surfaces of all the abdominal and pelvic organs. Fig. 10-11 shows how peritoneum lines the cavity's walls and the abdominal and pelvic organs in the coronal section of Fig. 10-10. Observe that the peritoneum encloses a cavity called the peritoneal cavity within the abdominopelvic cavity. In Fig. 10-11, the space of the peritoneal cavity is shaded light gray. Notice, in particular, that whereas all the abdominal and pelvic organs lie inside the abdominopelvic cavity, they lie, topologically, outside the peritoneal cavity.

12. What is the distinction between parietal peritoneum and visceral peritoneum?

The peritoneum that lines the inner surfaces of the abdominopelvic cavity's walls is called parietal peritoneum. The peritoneum that lines, or covers, the external surfaces of the abdominal and pelvic organs is called visceral peritoneum. In the abdominopelvic cavity, the parietal and visceral peritoneum form a continuous inner lining that surrounds the peritoneal cavity. The areas where parietal peritoneum is continuous with visceral

174

peritoneum are said to be areas of peritoneal reflection. For example, the area marked by an asterisk in Fig. 10-11 is an area of peritoneal reflection where visceral peritoneum covering the liver is continuous with parietal peritoneum lining the posterior abdominal wall.

13. What is the distinction between retroperitoneal and intraperitoneal organs?

A retroperitoneal organ is an abdominal organ whose anterior and side surfaces are covered by peritoneum but whose posterior surface is not covered by peritoneum and directly faces the posterior abdominal wall. The kidneys and adrenal glands are retroperitoneal organs. Fig. 10-11 displays the parts of the pancreas and duodenum that are retroperitoneal. However, these parts of the pancreas and duodenum are said more specifically to be secondarily retroperitoneal because they were were initially intraperitoneal during fetal development and became retroperitoneal secondarily only later during fetal development.

An intraperitoneal organ is an abdominal organ whose external surfaces are almost completely covered by peritoneum. The stomach, jejunum, ileum, and transverse colon are intraperitoneal organs.

Some abdominal organs, such as the liver, cannot be classified as retroperitoneal, secondarily retroperitoneal, or intraperitoneal organs. Observe that the liver's external surfaces are covered by visceral peritoneum except for a relatively large area called the bare area of the liver that is in direct contact with the undersurface of the diaphragm (Fig. 10-11). Pelvic organs, such as the urinary bladder and the rectum, are also in the category of organs that are not classified as retroperitoneal, secondarily retroperitoneal, or intraperitoneal organs.

14. What are peritoneal ligaments?

A peritoneal ligament is a double layer of peritoneum extending between two organs or an organ and an abdominal wall region. Peritoneal ligaments transmit the blood and lymphatic vessels and nerves that supply intraperitoneal organs. Fig. 10-12 shows a coronal section of four of the five largest peritoneal ligaments in the abdomen, all of which are shaded black. The peritoneal ligament which extends between the liver and the stomach is called the lesser omentum (LO). The peritoneal ligament which first descends from the stomach and then turns up to ascend to the transverse colon is called the greater omentum (GO). The peritoneal ligament which extends from the transverse colon to the pancreas and the posterior abdominal wall is called the transverse mesocolon (TM). Finally, the peritoneal ligament which extends from the posterior abdominal wall to the jejunum and ileum is called the mesentery of the small intestine (MSI). The fifth large peritoneal ligament extends from the posterior abdominal wall to the sigmoid colon and is called the sigmoid mesocolon.

15. What are the greater and lesser sacs of the peritoneal cavity, and where do they communicate with each other?

The peritoneal cavity can be divided into two unequal sub-cavities called the greater and lesser sacs (Fig. 10-13). The lesser sac, which may also be called the omental bursa, is the space within the peritoneal cavity that is bounded anteriorly by the lesser omentum, the stomach, and the descending part of the greater omentum. In Fig. 10-13, the lesser sac is shaded a darker gray than the greater sac. The remaining space within the peritoneal cavity is called the greater sac because its volume is greater than that of the lesser sac.

175

The lesser sac is sealed off superiorly by the liver and diaphragm, anteriorly by the lesser omentum, stomach, and descending part of the greater omentum, inferiorly by the ascending part of the greater omentum, the transverse colon, and the transverse mesocolon, and posteriorly by the upper posterior abdominal wall. On the left, it is sealed off by peritoneal ligaments. On the right, however, the lesser sac communicates with the greater sac through a passageway called the epiploic foramen (foramen of Winslow). The epiploic foramen is, in fact, the only passageway between the greater and lesser sacs.

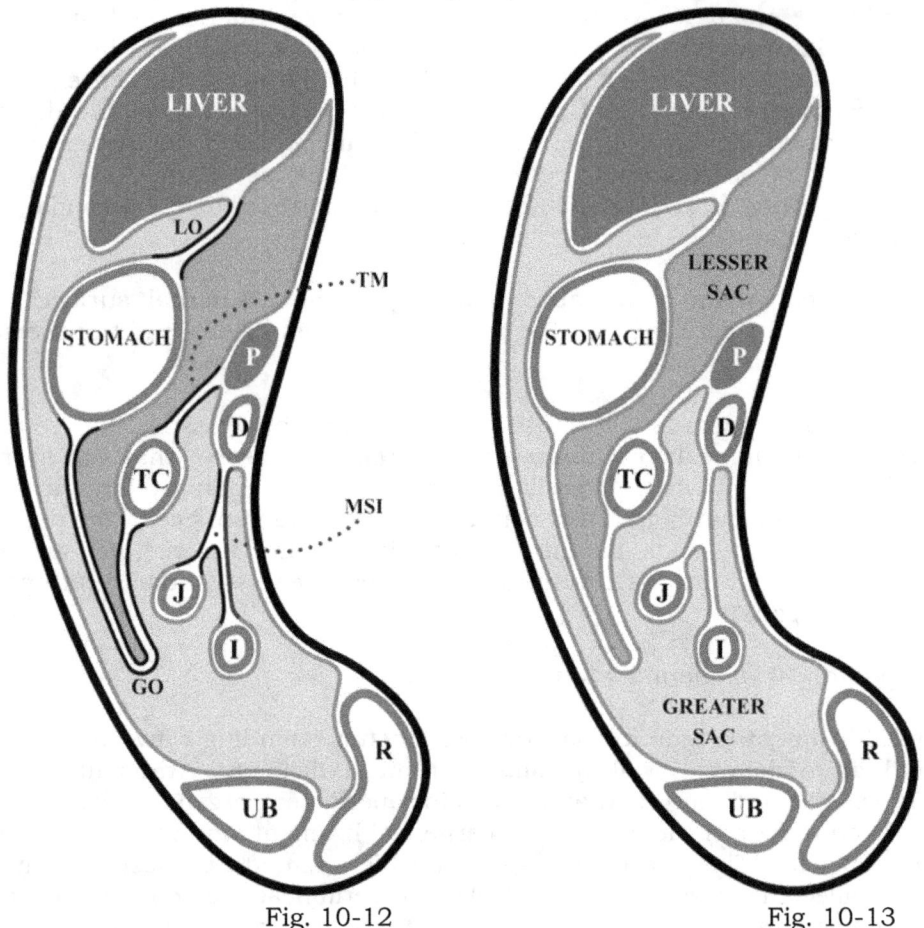

Fig. 10-12   Fig. 10-13

16. What are the borders of the epiploic foramen?

The epiploic foramen is bordered superiorly by the caudate lobe of the liver, anteriorly by the free, right border of the lesser omentum, posteriorly by the inferior vena cava, and inferiorly by the proximal half of the superior part of the duodenum. Three vessels extend within the lesser omentum near its free, right border: the common bile duct, hepatic artery proper, and the portal vein.

Regions of the Abdominal Cavity

When recording the results of a patient's physical examination, it is sometimes important to record the location of abdominal pain or an abdominal mass. Physicians use one of two methods to divide the abdomen into well-defined sub-regions. Both methods employ the mental projection of vertical and horizontal planes onto and through the anterior abdominal wall of the patient.

176

17. One of the methods divides the abdomen into four, roughly equal regions called quadrants. How is the abdominal cavity divided into quadrants?

The intersection of the median sagittal plane with the horizontal plane that passes through the umbilicus divides the abdomen into 4 quadrants: the right upper quadrant (RUQ), the left upper quadrant (LUQ), the left lower quadrant (LLQ), and the right lower quadrant (RLQ) (Fig. 10-14). The level of the umbilicus relative to the spine is not constant; its relative position varies with the position, age, and extent of obesity of a patient. In a lean patient lying in a supine position, the umbilicus generally lies at the level of the 3rd lumbar intervertebral disc, which is the intervertebral disc between the bodies of the 3rd and 4th lumbar vertebrae.

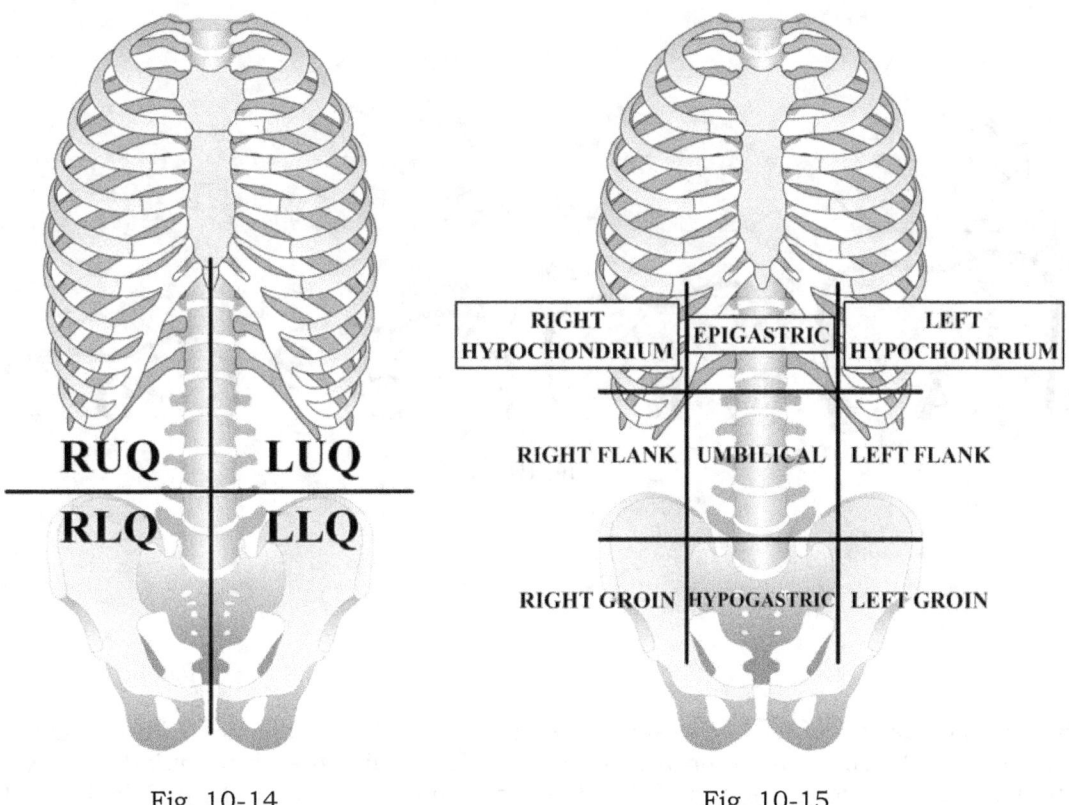

Fig. 10-14   Fig. 10-15

18. The other method divides the abdomen into nine, roughly equal regions. How is the abdominal cavity divided into the 9 roughly equal regions?

The intersections of the midclavicular planes with the transpyloric and intertubercular planes divide the abdomen into 9 regions: the right hypochondrium, epigastric, left hypochondrium, right flank, umbilical, left flank, right groin, hypogastric, and left groin regions (Fig. 10-15). Each midclavicular plane also passes through the midpoint between the anterior superior iliac spine and the pubic symphysis. The transpyloric plane, which lies midway between the superior borders of the manubrium of the sternum and the pubic symphysis, typically passes through the body of the 1st lumbar vertebra. The intertubercular plane passes anteriorly through the tubercles of the iliac crests of the coxal bones and posteriorly through the body of the 5th lumbar vertebra.

19. A person unfamiliar with the surface anatomy of the chest and abdomen might be surprised that each method for subdividing the abdomen places the highest abdominal regions deep to the lower parts of the rib cage. The diaphragm marks the highest level of the abdomen in the body. As a check on what you learned when you answered question 14 in the chapter on the chest, can you recall which of the four drawings in Fig. 10-16 most accurately represents the anterior surface projection of the diaphragm in a healthy adult lying supine on an examination table and breathing quietly?

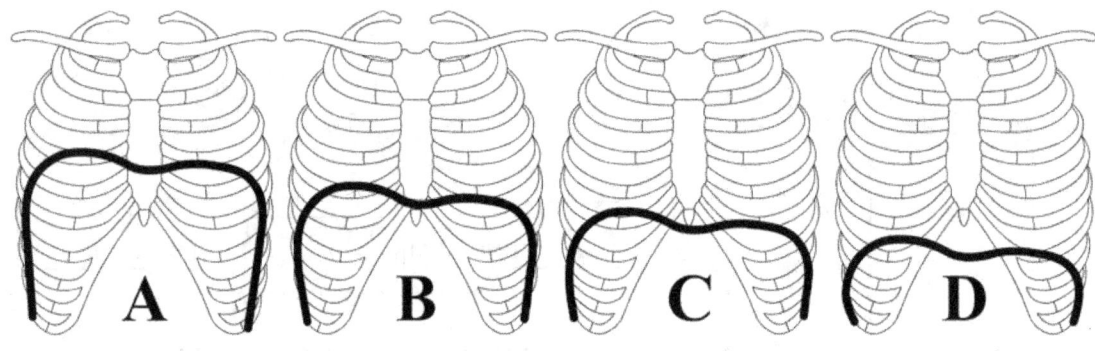

Fig. 10-16

Drawing B most accurately represents the anterior surface projection of the diaphragm in a healthy adult lying supine on an examination table and breathing quietly. Recall that, under these conditions, the top of the right dome intersects the right midclavicular line at about the level of the upper border of the right 5th rib, the top of the left dome intersects the left midclavicular line at about the level of the lower border of the left 5th rib, and the central tendon lies at the level of the xiphisternal joint.

Positions of Abdominal Viscera

The abdominal viscera which have the most constant positions in the abdominal cavity are the retroperitoneal and secondarily retroperitoneal viscera. Therefore, we will begin our examination of the positions of abdominal viscera relative to anterior and posterior abdominal wall landmarks by focusing on the location of (1) the retroperitoneal viscera (the abdominal aorta, inferior vena cava, kidneys and adrenal glands), (2) the secondarily retroperitoneal viscera (the ascending colon, descending colon, duodenum and pancreas), (3) the liver and gallbladder, and, finally, (4) the intraperitoneal organs (the spleen, stomach, cecum, appendix, transverse colon, and sigmoid colon).

20. The abdominal aorta and inferior vena cava are the largest abdominal blood vessels. Fig. 10-17 shows that the abdominal aorta begins at the level of the body of the 12th thoracic vertebra (where it passes through the aortic opening of the diaphragm) and then descends within the abdomen slightly to the left of the midline. Fig. 10-17 also shows that the inferior vena cava ascends within the abdomen slightly to the right of the midline and ends at the level of the body of the 8th thoracic vertebra (where it passes through the caval opening of the diaphragm). Indicate the vertebral level at which the abdominal aorta ends (upon bifurcating into the paired common iliac arteries) and the vertebral level at which the inferior vena cava begins (from the union of the paired common iliac veins).

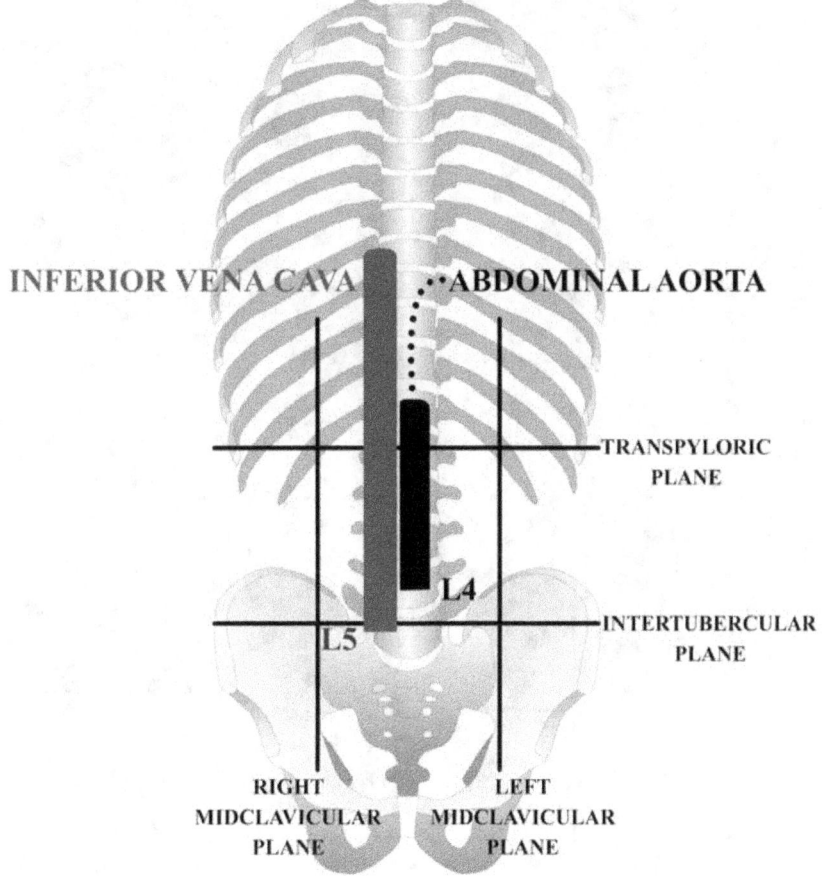

Fig. 10-17

The abdominal aorta ends at the level of the body of the 4th lumbar vertebra, which is at the level of the highest point of the iliac crest of the coxal bone. Notice that the abdominal aorta ends immediately above the lower border of the umbilical region of the abdomen. Deep palpation of the anterior abdominal wall in lean or minimally obese patients usually permits identification of aortic pulsations.

The inferior vena cava begins at the level of the body of the 5th lumbar vertebra. The inferior vena cava thus begins at the level of the lower border of the umbilical region.

179

21. The kidneys are the largest retroperitoneal organs. They lie nestled in the paravertebral gutters, which are the gutters alongside the spine. The right kidney lies at a level slightly lower than that of the left kidney because of the prominence of the liver on the right side of the body. Which of the 6 drawings in Fig. 10-18 most accurately represents the positions of the kidneys in a healthy adult who is lying in a supine position on an examination table and breathing quietly?

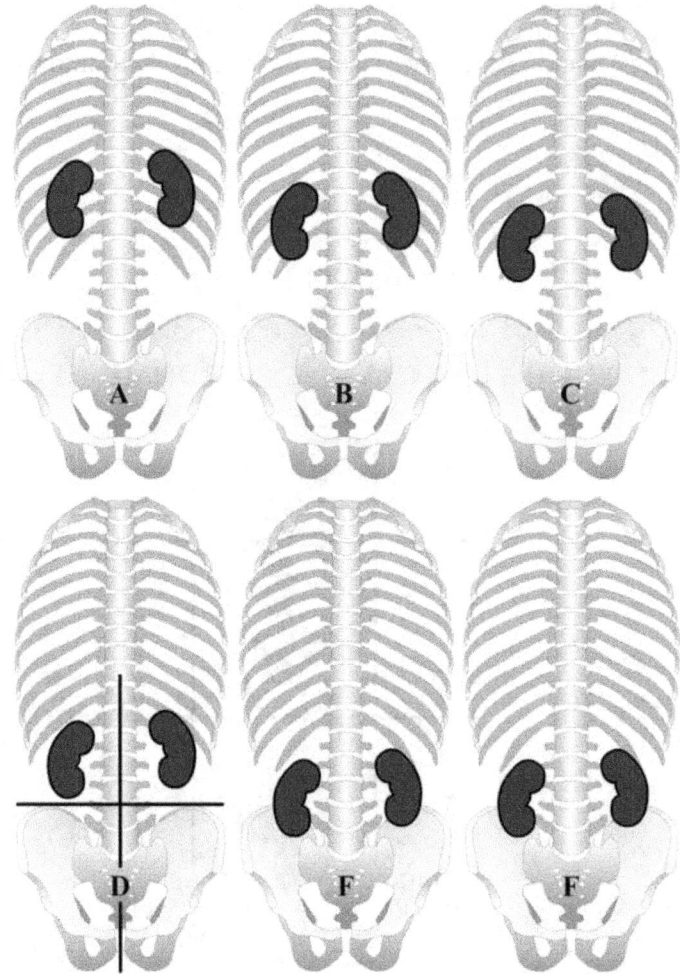

Fig. 10-18

Drawing D most accurately represents the positions of the kidneys in a healthy adult who is lying in a supine position on an examination table and breathing quietly. The left kidney's upper pole extends above the posterior segment of the left 11th rib, and the right kidney's upper pole extends above the posterior segment of the right 12th rib. Notice that the kidneys lie in the upper abdominal quadrants.

22. The adrenal glands are retroperitoneal viscera that have a constant spatial relationship relative to the kidneys. Describe this spatial relationship.

The adrenal glands lie directly atop the upper poles of the kidneys.

23. The largest secondarily retroperitoneal viscera of the abdomen are the ascending colon and descending colon. Describe their locations in the abdomen.

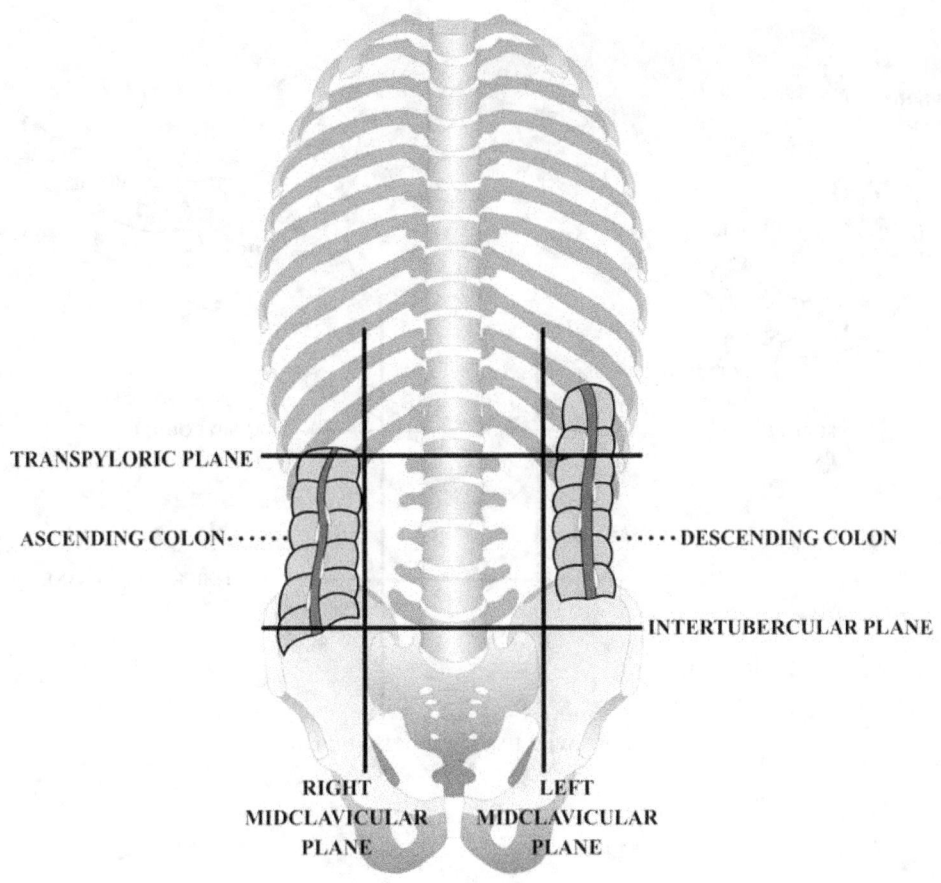

Fig. 10-19

Fig. 10-19 shows that the ascending colon and descending colon extend vertically through the flank regions of the abdomen. The ascending colon ascends from the right lower quadrant to the right upper quadrant; the descending colon descends from the left upper quadrant to the left lower quadrant.

24. Almost all the duodenum and almost all the pancreas are secondarily retroperitoneal. Describe their locations relative to each other and to the various subdivisions of the abdomen.

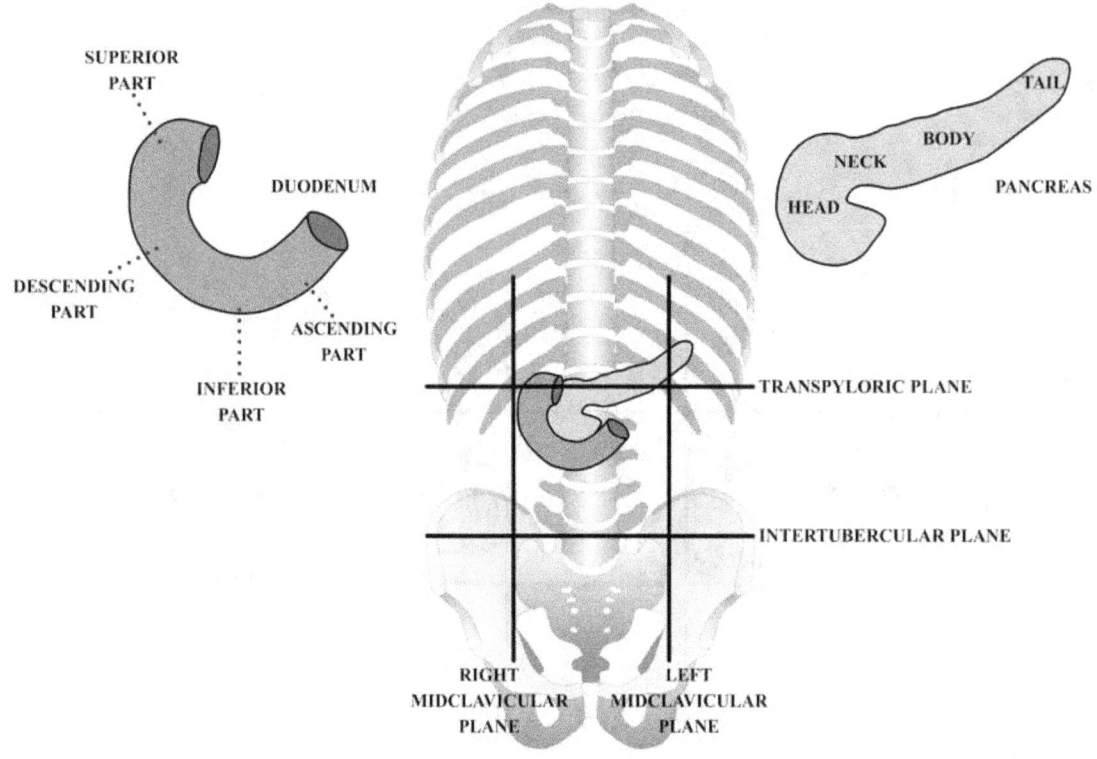

Fig. 10-20

Fig. 10-20 shows the parts of the duodenum and pancreas as well as their relationship to each other in the umbilical region of the abdomen. The duodenum is divisible into four parts called the superior, descending, inferior, and ascending parts. All the duodenum except for the proximal half of the superior part is secondarily retroperitoneal. The proximal half of the superior part of the duodenum is intraperitoneal; the peritoneum that covers its anterior and posterior surfaces is continuous superiorly with the free, right margin of the lesser omentum and inferiorly with the right margin of the greater omentum. The pancreas is also divisible into four parts called the head, neck, body, and tail. All the pancreas except for its tail is secondarily retroperitoneal. The tail is intraperitoneal and resides within the splenorenal ligament, a peritoneal ligament that extends from the left kidney to the spleen.

In the abdomen, the C-shaped duodenum curves around the head of the pancreas. The superior part of the duodenum and the neck of the pancreas lie at the level of the transpyloric plane, which is also the level of the body of the 1st lumbar vertebra. The inferior part of the duodenum lies at the level of the body of the 3rd lumbar vertebra. Almost all the duodenum and the head and neck of the pancreas lie within the umbilical region. The body of the pancreas ascends from the umbilical region to the epigastric region as it extends to the left; the tail of the pancreas ends in the left hypochondrium.

182

25. The liver and gallbladder have a constant relationship relative to each other and relatively constant relationships to the regions of the abdomen. Explain the anatomical basis of these relationships.

If an adult with a normal sized liver is lying in a supine position and breathing quietly, the liver resides almost completely in the right hypochondrium and epigastric regions (Fig. 10-21A). The liver's dome-shaped, upper surface is tethered superiorly against the undersurface of the diaphragm by a very short peritoneal ligament called the coronary ligament. Anteriorly, the liver is tethered close to the anterior abdominal wall by a short peritoneal ligament called the falciform ligament (Fig. 10-21B). The coronary and falciform ligaments together ensure that the liver moves up and down within the trunk of the body in tandem with the respiratory movements of the diaphragm.

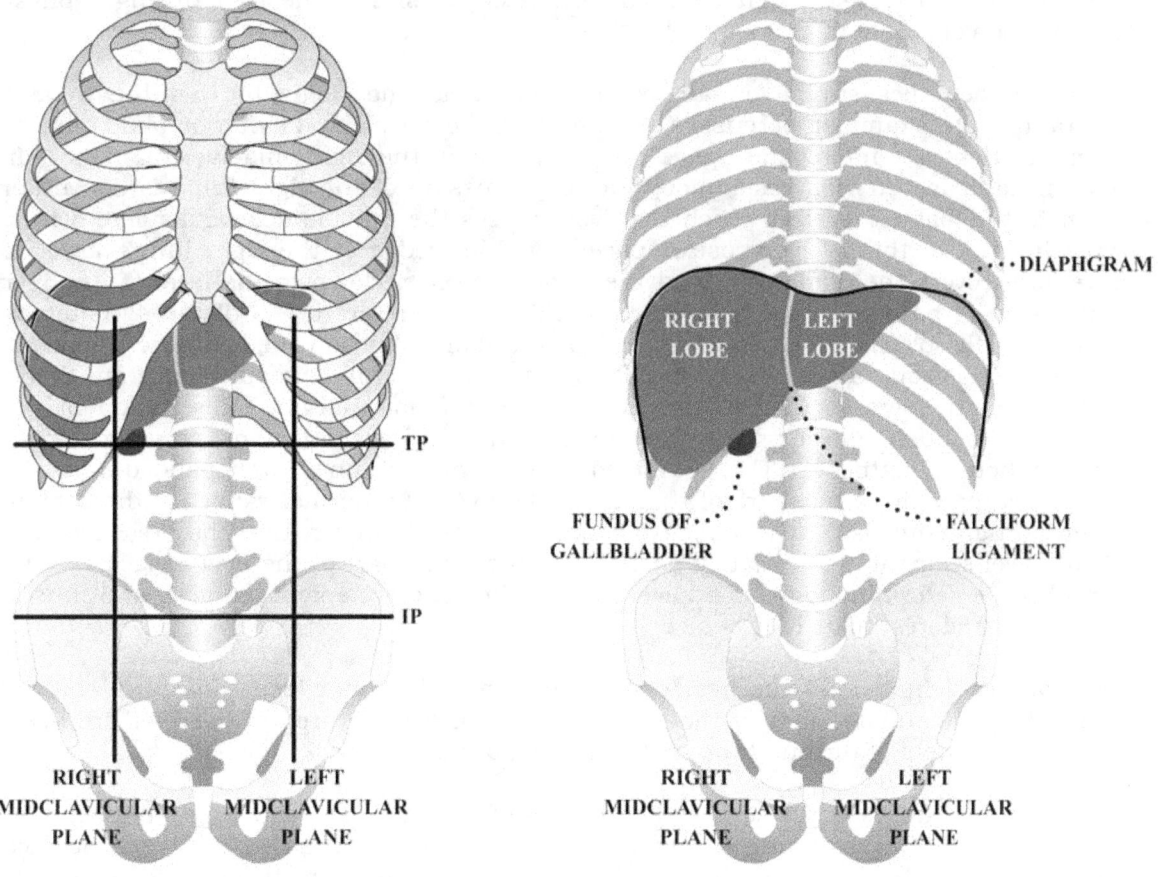

Fig. 10-21A: Relationship of the liver and the fundus of the gallbladder to the transpyloric plan (TP), intertubercular plane (IP), and midclavicular planes.

Fig. 10-21B: Relationship of the liver and the fundus of the gallbladder to the diaphragm and posterior rib cage.

The falciform ligament divides the liver anatomically into its right and left lobes (Fig. 10-21B). For a normal sized liver, the right lobe lies exclusively within the right upper quadrant, and the left lobe lies mostly in the left upper quadrant.

The gallbladder has a constant relationship with the liver because, in most individuals, the body of the gallbladder is in direct contact with the lower part of the liver's inferior surface, which is called the liver's visceral surface because it faces a variety of viscera. In most individuals, the neck of the gallbladder is attached to the liver's visceral surface via a very short peritoneal ligament, and the fundus of the gallbladder, which projects below the liver's lower margin, is completely covered with peritoneum. Observe that the fundus of the gallbladder typically crosses the transpyloric plane at or immediately to the left of the point where the transpyloric plane intersects the right costal margin.

26. Which part of the liver is palpable during a physical examination of a patient?

Deep palpation of the anterior abdominal wall generally permits palpation of the liver's anteroinferior margin. A normal liver's margin is soft, smooth, and non-tender; its course generally parallels the right costal margin of the rib cage.

27. Describe how percussion can be used to assess the size of the liver during a physical examination of a patient.

As the patient lies supine on an examination table and holds his/her breath at full expiration, the examiner percusses the anterior chest wall and anterior abdominal wall from the right 2nd intercostal space downward along the right midclavicular line. In a normal adult, percussion resonance (due to percussion of the right lung) is encountered down to the highest level at which the liver crosses the right midclavicular line (which is typically that of the 4th intercostal space). A thin wedge of the right lung's lower lobe anteriorly covers the diaphragm and the underlying liver down to the level of the 6th rib; percussion of both the right lung and the liver between the levels of the 4th intercostal space and the 6th rib produces a zone of percussion dullness called hepatic dullness. A zone of percussion flatness called hepatic flatness is generally encountered from the level of the 6th rib down to that of the liver's anteroinferior margin. Percussion of bowel segments inferior the liver produce percussion dullness, resonance, or tympany below the zone of hepatic flatness. The combined heights of the zones of hepatic dullness and hepatic flatness are a measure of the size of the liver; the normal range for the combined heights in an adult is 6 to 12 cm. It should be noted, however, that the presence of gas-filled bowel segments immediately posterior to the lower part of the liver's visceral surface can obscure determination of the lower limit of the zone of hepatic flatness, and thus lead to a faulty underestimate of liver size.

28. Part of the right lobe of the liver can be anatomically subdivided into lobes called the quadrate and caudate lobes of the liver. Which surface features of the liver are used to define the borders of the quadrate and caudate lobes of the liver?

The surface features which define the borders of the quadrate and caudate lobes are all features of the liver's visceral surface. The central area of the liver's visceral surface is called the porta hepatis because it is literally the gateway area through which the blood vessels that supply blood to the liver (namely, the left and right divisions of the portal vein and the left and right hepatic arteries) enter the liver (Fig. 10-22). The porta hepatis is also the gateway area by which the left and right hepatic ducts exit the liver. There are also four structures (two broad structures and two thin structures) that form impressions on the liver's visceral surface, and it is these four structures, in combination with the porta hepatis, that define the borders of the quadrate and caudate lobes (Fig. 10-22).

The two broad structures that form impressions on the liver's visceral surface are the terminal segment of the inferior vena cava and the body of the gallbladder (Fig. 10-22). The terminal segment of the inferior vena cava is the segment that passes through the caval opening of the diaphragm. The hepatic veins (which are the veins that receive all the venous blood drained from the liver) emerge from the bare area of the liver to unite with the terminal segment of the inferior vena cava. As noted in the answer to question 25, the body of the gallbladder in most persons lies in direct contact with the liver's visceral surface.

The two thin structures are postnatal cord-like remnants of fetal blood vessels. The ligamanetum venosum is the postnatal remnant of the fetal ductus venosus, and the ligamentum teres is the postnatal remnant of the fetal umbilical vein.

The inferior vena cava, gallbladder, ligamentum venosum, ligamentum teres, and porta hepatis all together form an **H**-shaped set of features on the liver's visceral surface. The crossbar and the upper parts of the vertical limbs of the **H** mark the borders of the caudate lobe, and the crossbar and the lower parts of the vertical limbs of the **H** mark the borders of the quadrate lobe.

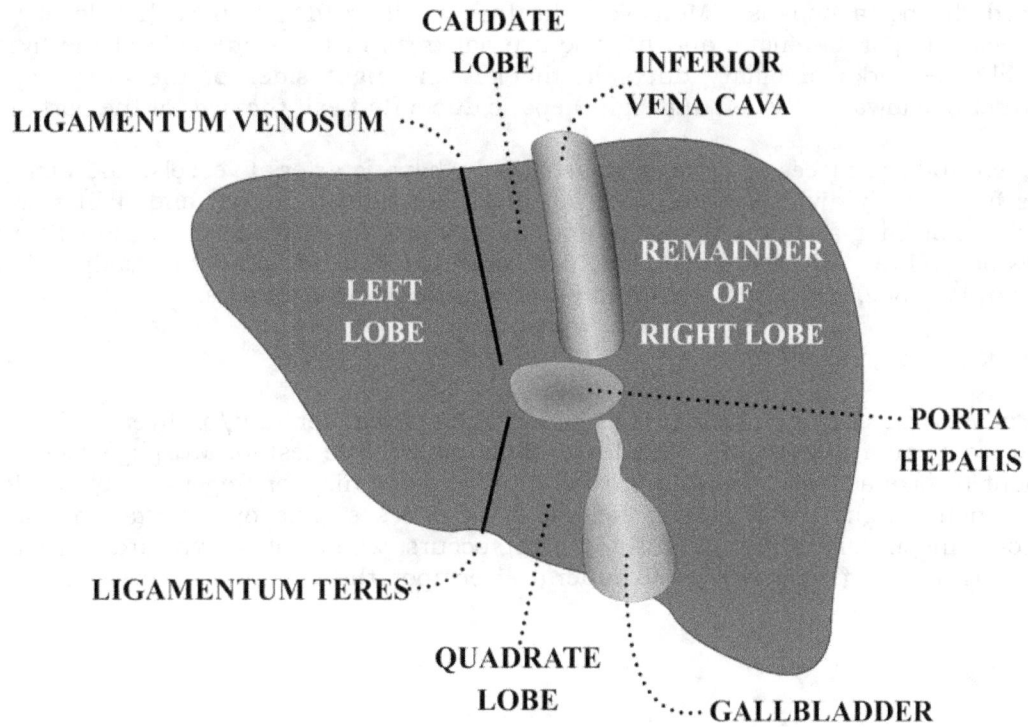

Fig. 10-22: Visceral surface of the liver.

29. What is the distinction between the anatomical division of the liver into right and left lobes by the falciform ligament and the functional division of the liver into right and left sides on the basis of its blood supply and bile production?

The liver is an exceptional organ in that it receives blood from two sources: the portal vein and the hepatic artery proper. The portal vein is an abdominal vein that receives all the venous blood drained from the spleen, pancreas, gallbladder, and the digestive tract from the lower third of the esophagus to the upper half of the anal canal. The portal vein arises behind the neck of the pancreas from the union of the splenic and superior mesenteric veins. The hepatic artery proper is an artery that arises indirectly from the celiac trunk. The portal vein and hepatic artery proper conduct, respectively, about 70% and 30% of the blood entering the liver.

At the porta hepatis, the portal vein divides into two branches called the left and right branches of the portal vein, and the hepatic artery proper divides into two branches called the left and right hepatic arteries. The left and right branches of the portal vein and the left and right hepatic arteries each give rise to a vascular tree (a tree-like network of vessels whose branches become progressively smaller as they extend throughout the parenchyma of the liver). The vascular tree that emanates from the left branch of the portal vein parallels the vascular tree that stems from the left hepatic artery, and the vascular tree that emanates from the right branch of the portal vein parallels the vascular tree that stems from the right hepatic artery. There is, therefore, on each side of the liver, a tree of portal venous vessels and a tree of hepatic arterial vessels that supply identical sectors of the liver in a parallel fashion.

There is also in each side of the liver a tree-like network of biliary ducts that conducts bile toward the porta hepatis. Moreover, the tree of biliary ducts in each side of the liver parallels the portal venous and hepatic arterial trees in the same side of the liver. The tree-like networks of biliary ducts in the left and right sides of the liver respectively conduct bile towards the left and right hepatic ducts that exit the liver at the porta hepatis.

The left and right trees of portal venous vessels, hepatic arterial vessels, and biliary ducts thus functionally divide the liver into left and right sides. The left and right sides of the liver do not have anatomical names and do not directly correspond to the left and right lobes of the liver. The left side of the liver consists of the left lobe, the quadrate lobe, and the caudate lobe. The remainder of the liver represents its right side.

30. What is Murphy's sign?

Murphy's sign demonstrates tenderness of the gallbladder and/or liver; it is generally regarded as an indication of gallbladder inflammation. The test for Murphy's sign is for the patient to take a deep breath as the examiner presses his/her fingers deeply underneath the patient's right costal margin (as the patient lies supine on an examination table). Sudden inspiratory arrest (Murphy's sign) occurs when the downward thrust of the diaphragm impinges a tender gallbladder or liver upon the examiner's fingertips.

31. The spleen is an intraperitoneal organ that has a relatively constant relationship to the rib cage. Explain the anatomical basis of this relationship, and identify which of the drawings in Fig. 10-23 most accurately represents the position of a normal sized spleen relative to the posterior rib cage.

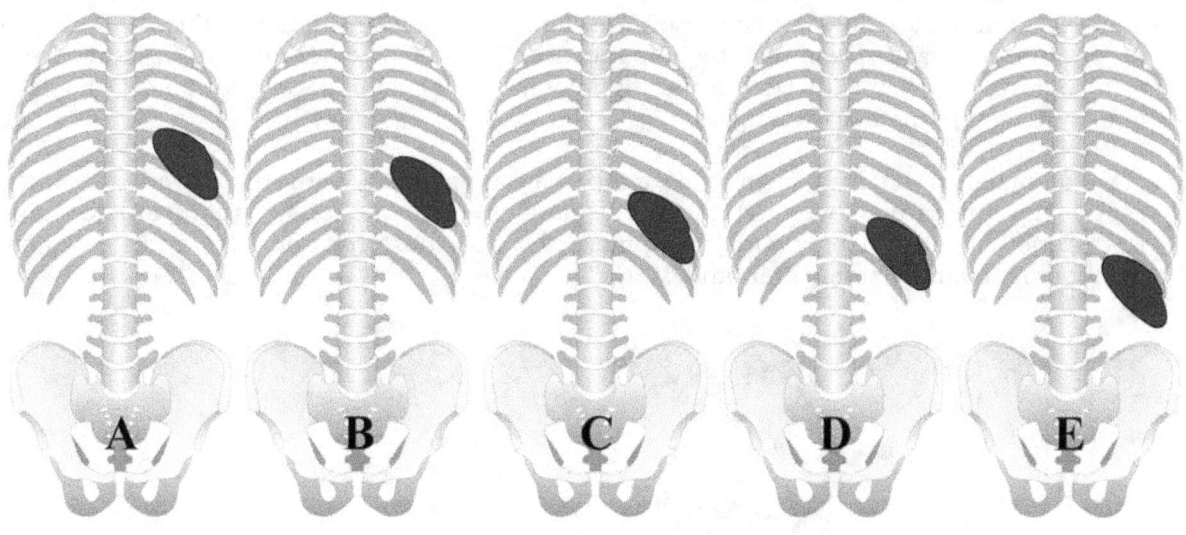

Fig. 10-23

Drawing C most accurately represents the position of a normal sized spleen relative to the posterior rib cage. The spleen lies directly deep to the lower posterolateral region of the left dome of the diaphragm. The spleen underlies the posterolateral segments of the left 9th, 10th, and 11th ribs, with its long axis paralleling the downward slope of the posterolateral segment of the left 10th rib.

The spleen's relatively constant relationship to the lower posterior part of the rib cage on the left side is due to the fact that it is attached to (a) the posterior abdominal wall by a relatively short peritoneal ligament called the splenorenal ligament, (b) the anterior abdominal wall by a relatively short peritoneal ligament called the phrenicocolic ligament, and (c) the upper greater curvature of the stomach by a relatively short peritoneal ligament called the gastrosplenic ligament.

The spleen is a large encapsulated organ whose parenchyma consists of dense vascular and lymphoid tissue. The spleen is always at risk of infarction or rupture by a severe blow to the lower left side of the rib cage. Splenic rupture can lead can lead to critical hypotension as a result of blood loss into the peritoneal cavity.

32. Is a normal sized spleen palpable during a physical examination of an adult patient?

A normal sized spleen is not palpable in most adults. However, in a small percentage of adults, the anteroinferior tip of a normal sized spleen may be palpable using the following technique: With the patient lying supine on an examination table and the examiner standing on the right side of the patient, the examiner uses his/her left hand to push upward (anteriorly) on the lower left side of the rib cage as his/her right hand presses gently downward (posteriorly) on the anterior abdominal wall at about the level of the umbilicus. As the patient breathes in deeply, the examiner slides the right hand upward on the anterior abdominal wall toward and under the left costal margin. As the patient's diaphragm descends during inspiration and forces the spleen inferiorly, the anteroinferior tip of a normal sized spleen may be palpable deep to the left costal margin. It is important to recognize, however, that such a finding may indicate splenomegaly (an enlarged spleen) and thus require further investigation.

33. The stomach is an intraperitoneal organ whose openings with the lower end of the esophagus (the cardiac opening) and the beginning of the duodenum (the pyloric opening) have constant relationships with the spine. Describe these relationships and the general position of an empty stomach in an adult.

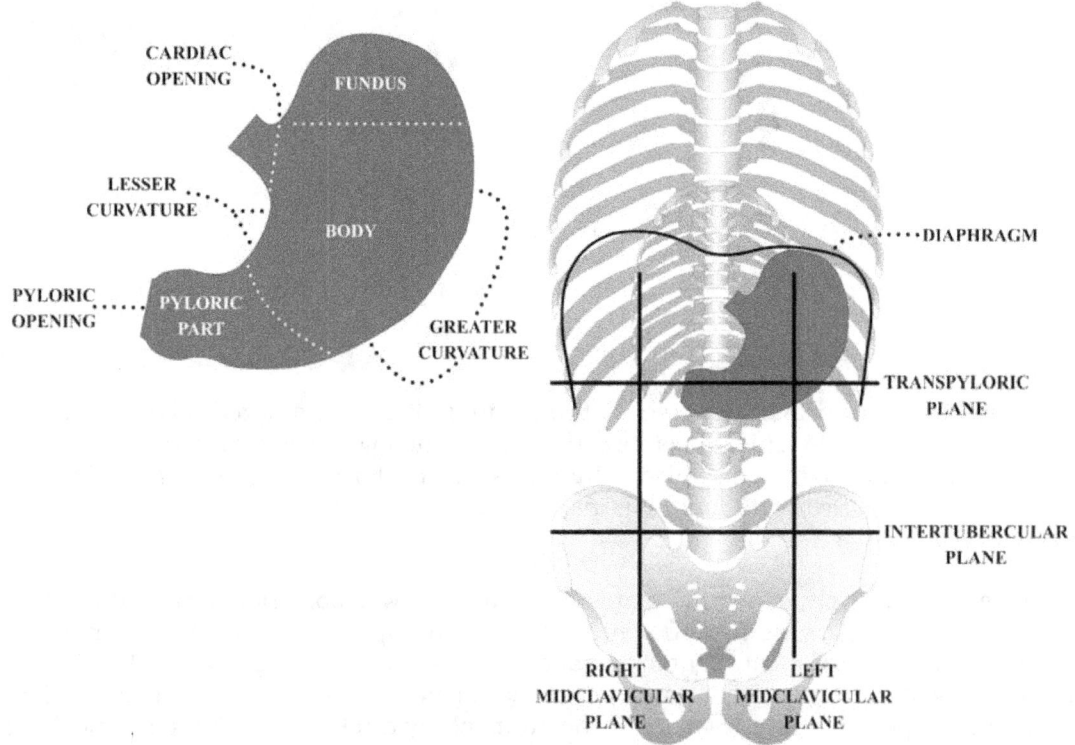

Fig. 10-24

Fig. 10-24 shows the general position of an empty stomach in an adult relative to the epigastric, umbilical and left hypochondrium regions of the abdomen. The cardiac opening typically lies at the level of the body of the 11th thoracic vertebra. In the anterior abdominal wall, the area immediately inferior to the tip of the xiphoid process lies just above the level of the body of the 11th thoracic vertebra. The pyloric opening lies at the level of the body of the 1st lumbar vertebra. Notice that the transpyloric plane owes its name to the fact that it marks the level of the pyloric opening.

188

The position and shape of the stomach are not fixed; they vary with the volume of the stomach contents, the position of the body, and the phases of respiration. The stomach is commonly found mainly in the left upper quadrant.

34. The cecum is an intraperitoneal segment of the large intestine. In which region of the abdomen does the cecum lie?

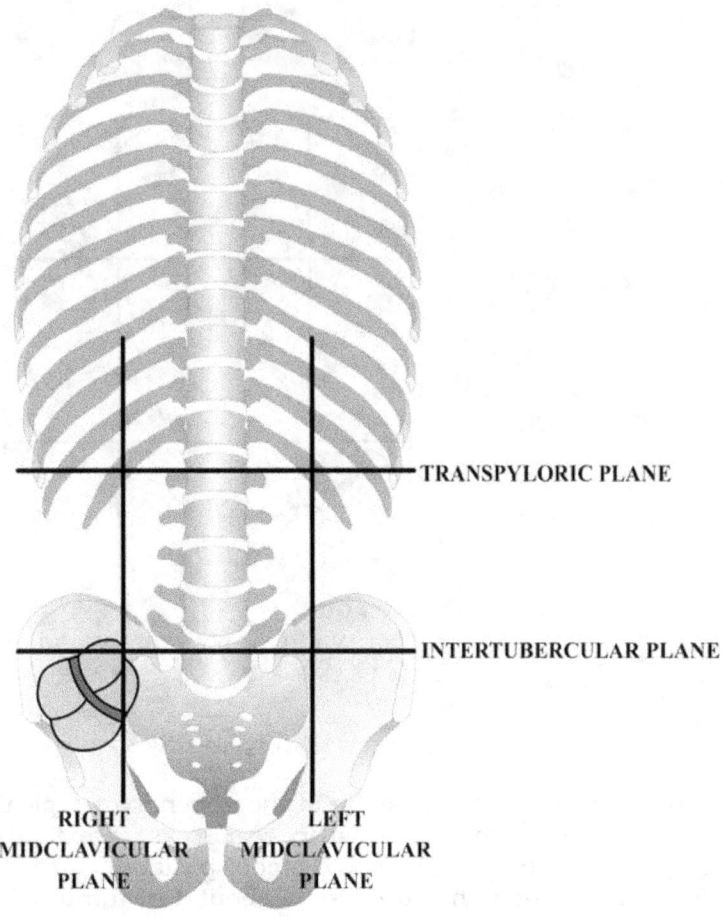

Fig. 10-25

Fig. 10-25 shows that the cecum lies in the right groin region. It exhibits a relatively constant position in the abdomen because it is commonly attached to the posterior abdominal wall by a very short peritoneal ligament.

35. The appendix is an intraperitoneal organ. Explain the anatomical basis of its variable position in the abdomen.

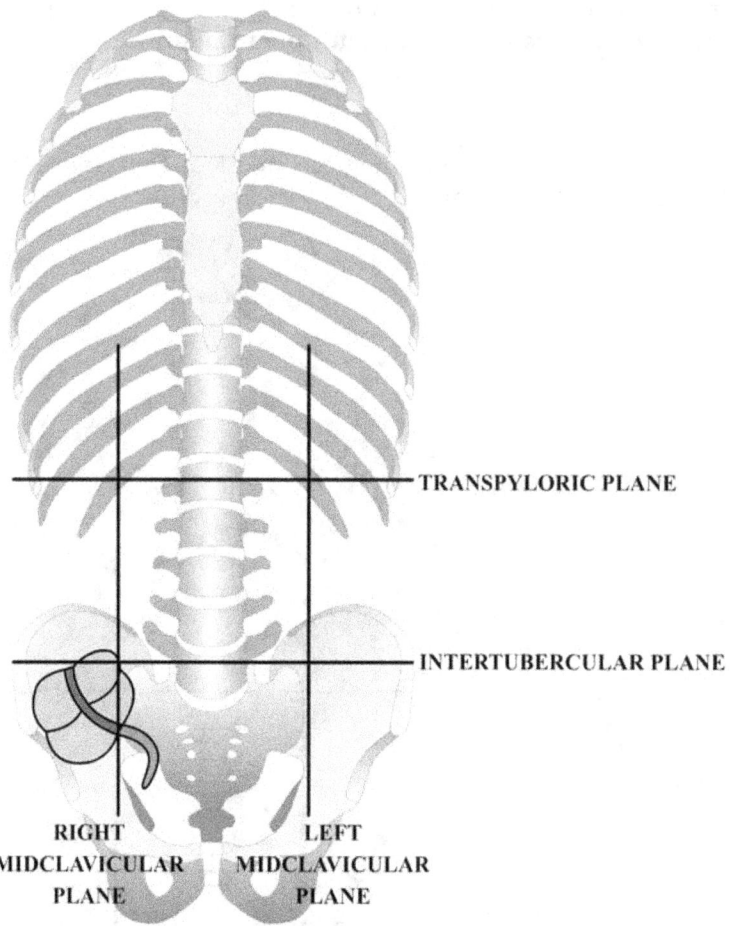

TRANSPYLORIC PLANE

INTERTUBERCULAR PLANE

RIGHT
MIDCLAVICULAR
PLANE

LEFT
MIDCLAVICULAR
PLANE

Fig. 10-26

Fig. 10-26 shows how the position of the appendix is typically portrayed in gross anatomy textbooks: lying in the hypogastric region as it extends inferiorly from the cecum. However, this is not the most common position of the appendix; the most common position, the position that occurs in about two-thirds of all persons, is the retrocecal position, the position in which the appendix lies behind the cecum. The appendix exhibits variable positions because it is attached to the mesentery of the small intestine by a relatively lax peritoneal ligament called the mesoappendix.

The part of the appendix which is most constant in position is its base; the base of the appendix is its open end, which is attached to the lower end of the cecum. The three teniae coli bands of the cecum converge at the base of the appendix.

36. The transverse colon and the sigmoid colon are both intraperitoneal segments of the large intestine. Describe their typical positions.

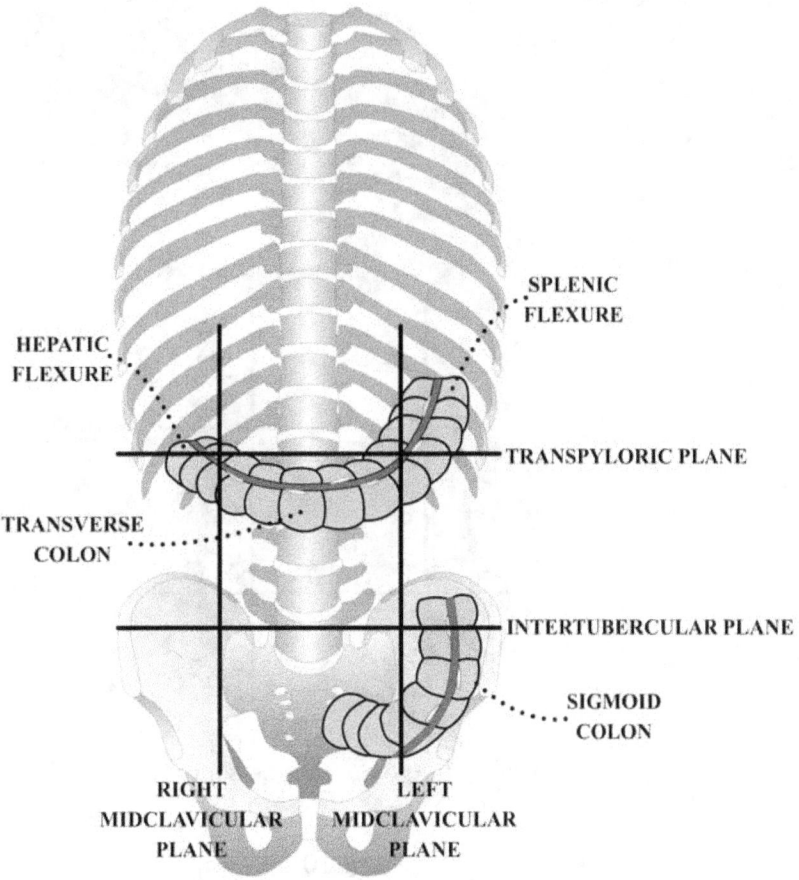

Fig. 10-27

Fig. 10-27 shows the typical positions of the transverse colon and sigmoid colon in the abdomen. As the transverse colon extends from the hepatic flexure to the splenic flecture, it typically traverses the umbilical region to end in the left hypochondrium. The sigmoid colon begins near the level of the intertubercular plane and then follows an inferomedial course as it descends through the left groin region into the lower part of the hypogastric region.

191

37. The jejunum and ileum are both intraperitoneal segments of the small intestine. Describe their typical positions.

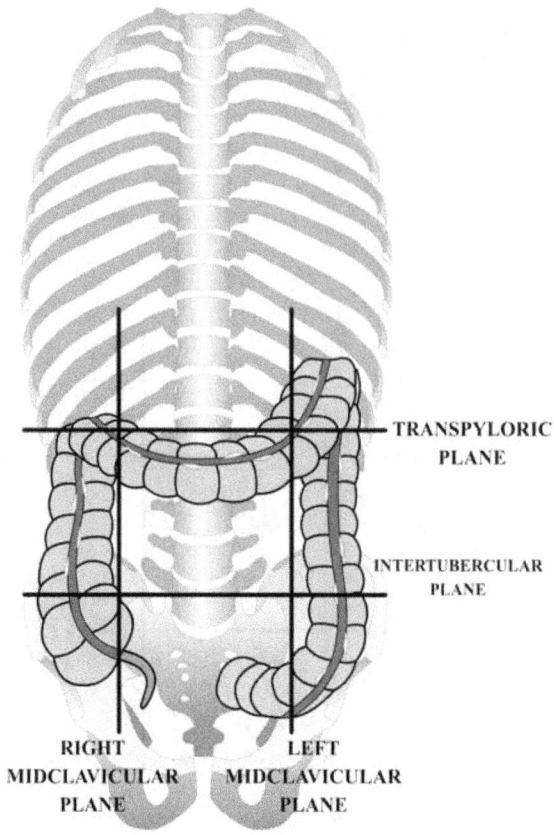

Fig. 10-28

Fig. 10-28 shows the course that the segments of the large intestine take as they first ascend in the right side of the abdomen, next extend to the left from the hepatic flexure, then descend in the left side of the abdomen, and finally follow an S-shaped course toward the pelvis. The central area of the abdomen bounded by these large intestine segments consists largely of the umbilical and hypogastric regions, and it is in this central area that most of the coils of the jejunum and ileum typically reside.

The jejunum represents roughly the proximal two-fifths of the small bowel coils. The major distinctions between the jejunum and ileum are as follows: (1) The jejunal branches of the superior mesenteric artery form fewer tiers of arterial arches (arterial anastomoses) in the mesentery of the small intestine than do the ileal branches. Straight arteries called vasa recta arise form the most distal tiers to extend straight toward the small bowel segment they supply. (2) The vascularity of the jejunal wall is denser than that of the ileal wall. (3) The plicae circulares of the jejunal mucosa are thicker and more numerous than those of the ileal mucosa. This difference contributes to the greater thickness of the jejunal wall; palpation of wall thickness can be used during surgery to distinguish upper jejunal coils from lower ileal coils. (4) Peyer's patches are less numerous in the jejunum than in the ileum. Peyer's patches are aggregates of lymphoid tissue distributed along the antimesenteric border of the jejunum and ileum. (5) The mesentery of the jejunum bears less fat than the mesentery of the ileum.

192

38. Disease or injury of an abdominal organ or gland can produce three types of pain: visceral pain, referred pain, and/or somatic pain. What is visceral pain?

Visceral pain is a dull, sickening pain which is poorly localized to one of the midline regions (epigastric, umbilical, or hypogastric) of the abdomen. Visceral pain is produced by the stimulation of visceral pain fibers. Visceral pain fibers are sensitive to acute stretching and anoxia. Visceral pain fiber endings are located in the muscular walls of hollow abdominal organs (such as the walls of the small and large bowel, the extrahepatic biliary ducts, the ureters, and the bladder) and in the fibrous capsules of solid abdominal organs (such as the capsules of the liver, spleen, and kidneys).

39. Which abdominal viscera, when diseased or injured, typically produce visceral pain in (a) the epigastric region, (b) the umbilical region, and (c) the hypogastric region?

Disease or injury of the abdominal viscera supplied by branches of the celiac artery typically produces visceral pain in the epigastric region. These viscera include the lower third of the esophagus, the stomach, the duodenum, the pancreas, the spleen, the liver, and the gallbladder. Each of these viscera is supplied by pain fibers that reach the organ or gland by coursing alongside the branches of the celiac artery that supply the organ or gland.

Disease or injury of the abdominal viscera supplied by branches of the superior mesenteric artery typically produces visceral pain in the umbilical region. These viscera include the pancreas; the duodenum, jejunum and ileum; and the appendix, cecum, ascending colon, and transverse colon. Each of these viscera is supplied by pain fibers that reach the organ or gland by coursing alongside the branches of the superior mesenteric artery that supply the organ or gland. Observe that the pancreas and duodenum are each supplied by branches of both the celiac and superior mesenteric arteries, and thus can elicit visceral pain in the epigastric and/or umbilical regions.

Disease or injury of the abdominal viscera supplied by branches of the inferior mesnteric artery typically produces visceral pain in the hypogastric region. These viscera include the transverse colon, descending colon, sigmoid colon, rectum, and the upper half of the anal canal. Each of these viscera is supplied by pain fibers that reach the organ by coursing alongside the branches of the inferior mesenteric artery that supply the organ. Observe that the transverse colon is supplied by branches of both the superior and inferior mesenteric arteries, and thus can elicit pain in the umbilical and/or hypogatric regions.

40. What is referred pain?

Referred pain is pain that a diseased or injured organ or gland refers to a cutaneous region of the body. The dermatomes of the painful cutaneous region are innervated by the same spinal cord segments that innervate the diseased or injured viscus. The painful cutaneous region is generally anatomically distant from the site of the diseased or injured viscus.

41. What is somatic pain?

Somatic pain is produced by the stimulation of parietal peritoneum pain fibers. Inflammation of the parietal peritoneum of the anterior abdominal wall produces somatic pain that is sharp and limited to the region of inflammation. By contrast, inflammation of the parietal peritoneum of the posterior abdominal wall produces somatic pain that is diffuse and localized to the midline. A diseased abdominal organ can produce somatic pain when its inflammatory process extends to the parietal peritoneum of those abdominal wall regions with which the organ contacts.

Stretching exacerbates the somatic pain of inflamed parietal peritoneum. Minimally inflamed parietal peritoneum in an anterior abdominal wall region causes an individual to voluntarily guard, or split (stiffen), the inflamed region against unwanted movement. Worsening of the inflammation ultimately elicits reflex contraction and rigidity of the abdominal wall muscles overlying the inflamed region.

42. What is rebound tenderness?

The sensitivity of inflamed parietal peritoneum to sudden stretching can be detected by pressing the adducted fingers of one hand into a region of the anterior abdominal wall and then suddenly withdrawing the fingers. The wave of sudden rebound movement that spreads throughout the abdominal walls elicits tenderness called rebound tenderness in those wall regions lined by inflamed parietal peritoneum. It is preferable to apply the finger pressure at a site distant from suspected sites of inflammation. An examiner must use discretion in applying appropriate pressure during testing for rebound tenderness, as the pain can be quite severe.

Dissection of the pelvis and perineum in gross lab focuses on identification of
  (1) the skeletal, cartilaginous, and ligamentous boundaries of the pelvic inlet
       and pelvic outlet,
  (2) the pelvic viscera and their blood supply in both sexes,
  (3) piriformis, obturator internus, levator ani, and coccygeus,
  (4) the anal canal and its blood supply and venous drainage,
  (5) the sacral plexus and pudendal nerve,
  (6) the skeletal muscles of the perineum,
  (7) the erectile tissues of the penis and clitoris, and
  (8) the contents of the scrotum.
Four topics dominate the pelvis and perineum anatomy most frequently applied in clinical practice: (1) bimanual examination of the female pelvis, (2) obstetrical pelvic measurements, (3) rectal exam of the prostate, and (4) physical examination of the scrotum.

Pelvis

1. Describe the the structures that border the pelvic inlet anteriorly, laterally, and posteriorly.

2. In instances of urethral obstruction, urine can be drained from a swollen urinary bladder in an adult by inserting a needle into the suprapubic region of the anterior abdominal wall. Explain why in this procedure, which is called suprapubic cystotomy, the needle does not pass through the peritoneal cavity.

3. In the female pelvis, which structures is the vagina in direct contact with (a) anteriorly and (b) posteriorly?

4. Describe the lowest region of the peritoneal cavity in the female cavity.

5. Which fornix of the vaginal vault underlies the pouch of Douglas?
   _____ Anterior fornix
   _____ Left lateral fornix
   _____ Posterior fornix
   _____ Right lateral fornix

6. In the female pelvis, the longitudinal axis of the body of the uterus can be bent either forward or backward relative to the longitudinal axis of the vagina. Furthermore, the body of the uterus can also be bent either forward or backward relative to its cervix. What terms are used to describe the orientation of the body of the uterus relative to the vagina and the cervix of the uterus?

7. What physical properties of the uterus can typically be evaluated during a pelvic exam?

8. In anticipation of a vaginal delivery of an infant, it is important to measure the narrowest diameter of the birth canal at the pelvic inlet and the midregion of the pelvis. Describe the narrowest diameter at the pelvic inlet and the mid pelvis.

9. Which nerve innervates all the perineal muscles and the skin of the perineum? Which spinal nerves provide nerve fibers for this nerve?

10. Which bony pelvis landmark is used to locate the site at which anesthetic is infiltrated during a pudendal nerve block?

10. Which normal structures on each side are palpable during physical examination of the scrotum?

11. Describe the difference between internal and external hemorrhoids.

END OF QUESTIONS IN PART A OF THE CHAPTER ON THE PELVIS AND PERINEUM

PELVIS AND PERINEUM - Part A: Questions and Answers

Dissection of the pelvis and perineum in gross lab focuses on identification of
  (1) the skeletal, cartilaginous, and ligamentous boundaries of the pelvic inlet
      and pelvic outlet,
  (2) the pelvic viscera and their blood supply in both sexes,
  (3) piriformis, obturator internus, levator ani, and coccygeus,
  (4) the anal canal and its blood supply and venous drainage,
  (5) the sacral plexus and pudendal nerve,
  (6) the skeletal muscles of the perineum,
  (7) the erectile tissues of the penis and clitoris, and
  (8) the contents of the scrotum.

Four topics dominate the pelvis and perineum anatomy most frequently applied in clinical practice: (1) bimanual examination of the female pelvis, (2) obstetrical pelvic measurements, (3) rectal exam of the prostate, and (4) physical examination of the scrotum.

Pelvis

1. Describe the the structures that border the pelvic inlet anteriorly, laterally, and posteriorly.

   The superior opening of the pelvis is called the pelvic inlet. The border of the pelvic inlet is called the pelvic brim, and it is formed anteriorly by the upper border of the pubic symphysis and the pubic crests of the coxal bones, laterally by the iliopectineal lines of the coxal bones, and posteriorly by the sacral promontory (the sacral promontory is the anterosuperior edge of the body of the 1st sacral vertebra).

2. In instances of urethral obstruction, urine can be drained from a swollen urinary bladder in an adult by inserting a needle into the suprapubic region of the anterior abdominal wall. Explain why in this procedure, which is called suprapubic cystotomy, the needle does not pass through the peritoneal cavity.

   The only surface of the urinary bladder that is covered with peritoneum is the bladder's superior surface. In an adult, the bladder (when empty) lies completely in the pelvic cavity, immediately posterior to the upper border of the pubic symphysis and the pubic parts of the coxal bones. Accordingly, in an adult with an empty bladder, the bladder's superior surface lies immediately below the pelvic inlet.

   The bladder accommodates increases in urine volume through primarily distention of its superior surface. As urine accumulates in the bladder, the bladder's superior surface bulges upward and becomes more rounded. In an adult, this upward bulging commonly elevates the bladder's superior surface above the pelvic inlet. Continuing expansion of the bladder's superior surface ultimately lifts peritoneum from the posterior surface of the anterior abdominal wall onto the bladder's bulging, superior surface; this change brings the posterior surface of the anterior abdominal wall and the bladder's superior surface into direct contact with each other above the pubic symphysis. This direct contact makes it possible to insert a needle into the interior of the bladder from the suprapubic, anterior abdominal wall without passing through the peritoneal cavity.

3. In the female pelvis, which structures is the vagina in direct contact with (a) anteriorly and (b) posteriorly?

Anteriorly, the vagina is in direct contact with the urethra and the base of the bladder. Posteriorly, the vagina is in direct contact with the perineal body and the lower third of the rectum.

4. Describe the lowest region of the peritoneal cavity in the female cavity.

In the midline region of the female pelvis, the peritoneum which covers the anterior surface of the middle third of the rectum is continuous inferiorly with the peritoneum that covers the uppermost posterior surface of the vagina. This peritoneal reflection from the rectum onto the vagina lines the lowest part of a pouch in the floor of the female pelvis called the pouch of Douglas (rectouterine pouch).

5. Which fornix of the vaginal vault underlies the pouch of Douglas?
   _____ Anterior fornix
   _____ Left lateral fornix
   __x__ Posterior fornix
   _____ Right lateral fornix

6. In the female pelvis, the longitudinal axis of the body of the uterus can be bent either forward or backward relative to the longitudinal axis of the vagina. Furthermore, the body of the uterus can also be bent either forward or backward relative to its cervix. What terms are used to describe the orientation of the body of the uterus relative to the vagina and the cervix of the uterus?

A uterus is said to be anteverted if its longitudinal axis is bent forward relative to the longitudinal axis of the vagina; the uterus is said to be retroverted if its longitudinal axis is bent backward relative to the longitudinal axis of the vagina.

A uterus is said to be anteflexed if its body is bent forward relative to its cervix; the uterus is said to be retroflexed if its body is bent backward relative to the cervix.

7. What physical properties of the uterus can typically be evaluated during a pelvic exam?

The pelvic exam typically provides for evaluation of the uterus' position, size, shape, consistency, and mobility.

8. In anticipation of a vaginal delivery of an infant, it is important to measure the narrowest diameter of the birth canal at the pelvic inlet and the midregion of the pelvis. Describe the narrowest diameter at the pelvic inlet and the mid pelvis.

The narrowest diameter near the pelvic inlet is an anteroposterior diameter called the obstetric conjugate diameter; it is the distance from the midpoint of the sacral promontory to the closest point on the posterior surface of the pubic symphysis. The obstetric conjugate diameter cannot be measured directly by clinical assessment.

The narrowest diameter in the bony pelvis is a transverse diameter that lies at the midlevel of the pelvic cavity. It is called the interspinous diameter because it is the distance between the ischial spines. The interspinous diameter is the transverse diameter of a plane called the plane of least pelvic dimensions. The inferior margin of the pubic symphysis marks the most anterior border of the plane of least pelvic dimensions. The anteroposterior diameter of the plane of least pelvic dimensions is the shortest anteroposterior diameter in the bony pelvis.

Perineum

9. Which nerve innervates all the perineal muscles and the skin of the perineum? Which spinal nerves provide nerve fibers for this nerve?

On each side, the pudendal nerve innervates all the perineal muscles (in particular, the external anal sphincter and the skeletal muscles that control the timing of micturition) and the skin of the perineum. S2, S3, and S4 provide all the nerve fibers for the pudendal nerve.

10. Which bony pelvis landmark is used to locate the site at which anesthetic is infiltrated during a pudendal nerve block?

Ischial spine.

10. Which normal structures on each side are palpable during physical examination of the scrotum?

Testis, epididymis, and vas deferens.

11. Describe the difference between internal and external hemorrhoids.

Internal hemorrhoids are varicosities of the internal rectal venous plexus in the upper half of the anal canal. Internal hemorrhoids are lined by mucosa typical of that of the rectum; the mucosa is insensitive to pain. External hemorrhoids are varicosities of the external rectal venous plexus in the lower half of the anal canal. External hemorrhoids are lined by skin, and thus are sensitive to pain. The pectinate line, which lies at the level of the anal valves, marks the boundary between the upper and lower halves of the anal canal.

Dissection of the head & neck in gross lab focuses on identification of
  (1) the nerves, blood vessels, muscles, and glands of the head and neck,
  (2) the bones of the skull, the hyoid bone, and the cartilages of the larynx,
  (3) the dural infoldings and dural venous sinuses of the cranial cavity, and,
  (4) the cranial nerves as they exit and enter the base of the skull.

The head & neck anatomy most frequently applied in clinical practice includes the testing of cranial nerve function; palpation of the carotid pulse, thyroid gland, and head & neck lymph nodes; and the assessment of jugular venous pressure.

Neck

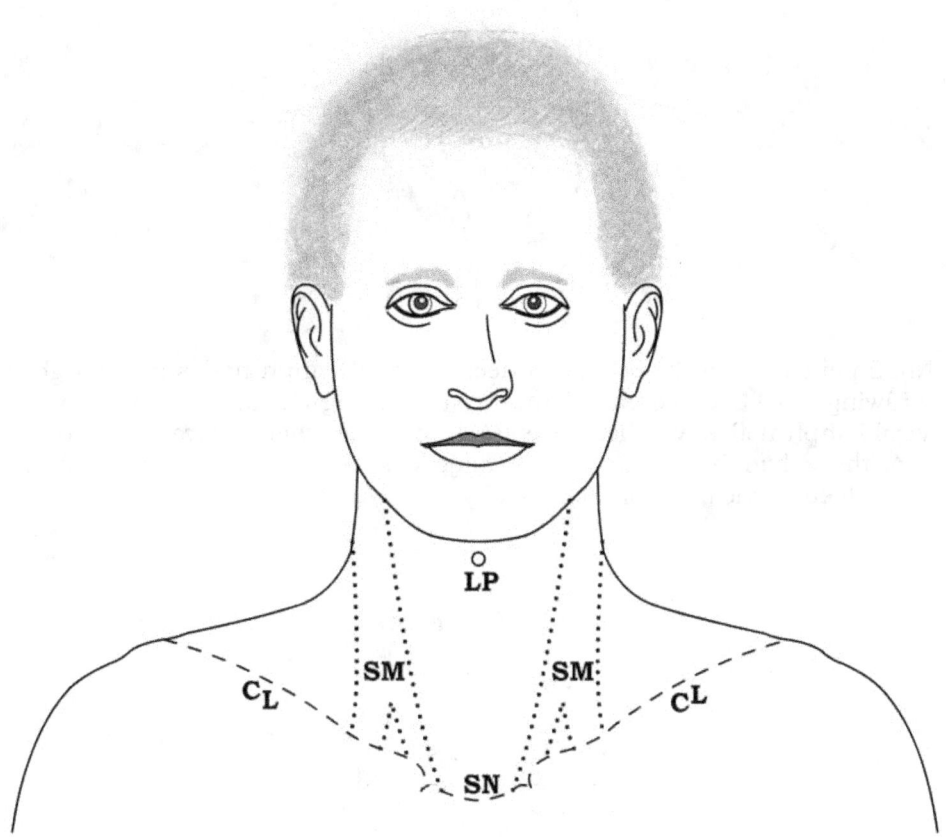

Fig. 12-1

1. Fig. 12-1 illustrates the surface landmarks that can generally be identified upon visual inspection of a patient's neck, namely, the laryngeal prominence (Adam's apple) (LP), the sternocleidomastoid muscles (SM), the suprasternal notch of the manubrium of the sternum (SN), and the upper margins of the clavicles (CL). Using a No. 2 pencil, shade in the anterior and posterior triangles of the neck on the right side of the patient's neck. Describe the boundaries of both triangles.

2. Which cranial nerve descends across the floor of the posterior triangle of the neck, and why is it susceptible to injury by relatively superficial neck lacerations in the posterior triangle of the neck?

3. Using a No. 2 pencil, circumscribe the area in Fig. 12-2 where the pulse of the right common carotid artery can be palpated in the neck.

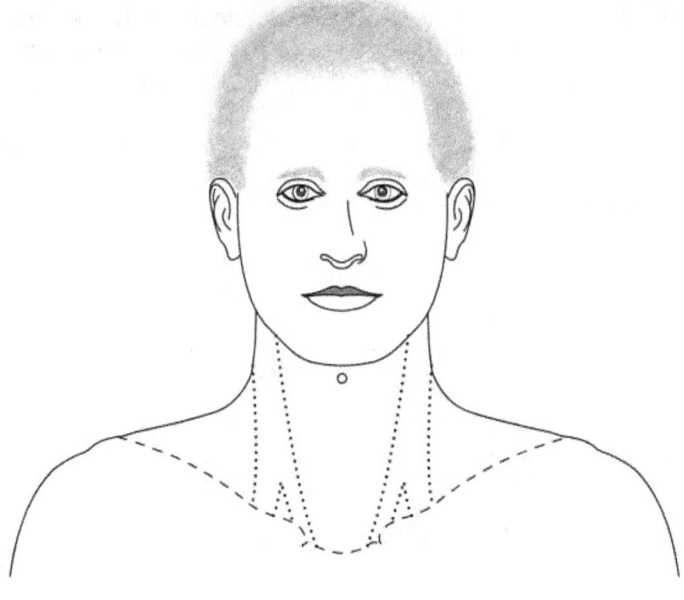

Fig. 12-2

4. Using a No. 2 pencil, trace the course of deep cervical lymph nodes in the right side of the neck by drawing 7-10 deep cervical lymph nodes in Fig. 12-3. Mark with an asterisk the deep cervical lymph nodes which are easiest to palpate when enlarged; discuss the clinical relevance of these lymph nodes. Which major blood vessel of the neck is most closely related to the deep cervical lymph nodes?

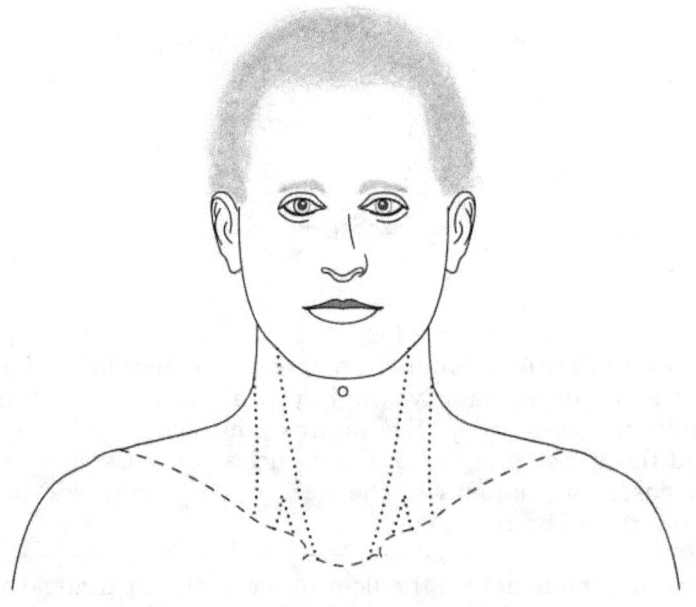

Fig. 12-3

5. Using a No. 2 pencil, draw 3 lymph nodes to mark the location of submental lymph nodes and 3 lymph nodes to mark the location of submandibular lymph nodes in Fig. 12-4. Describe the tissues from which the submental and submandibular nodes receive lymph.

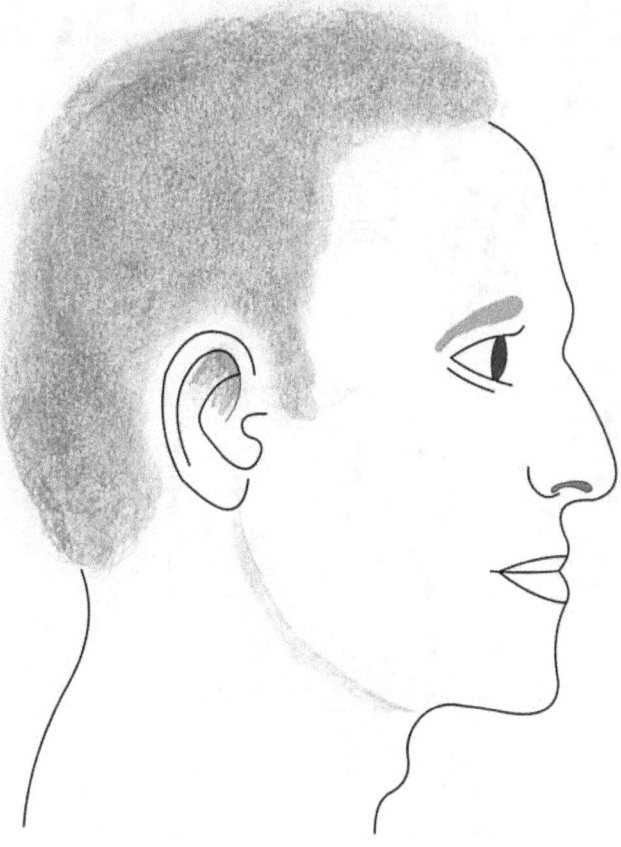

Fig. 12-4

6. Which surface features of the neck are used to locate and palpate the lateral lobes of the thyroid gland?

7. If a swelling is palpated in the area overlying one of the lateral lobes, what physical exam maneuver can be employed to investigate the likelihood that the swelling is associated with the thyroid gland?

8. How far superiorly and inferiorly can diffuse enlargements of the lateral lobes of the thyroid gland (diffuse goiters) extend?

9. The thick, dark lines in Fig. 12-6 represent a blood vessel that typically can be seen on each side of the neck superficial to the sternoclediomastoid muscle. Identify the blood vessel.

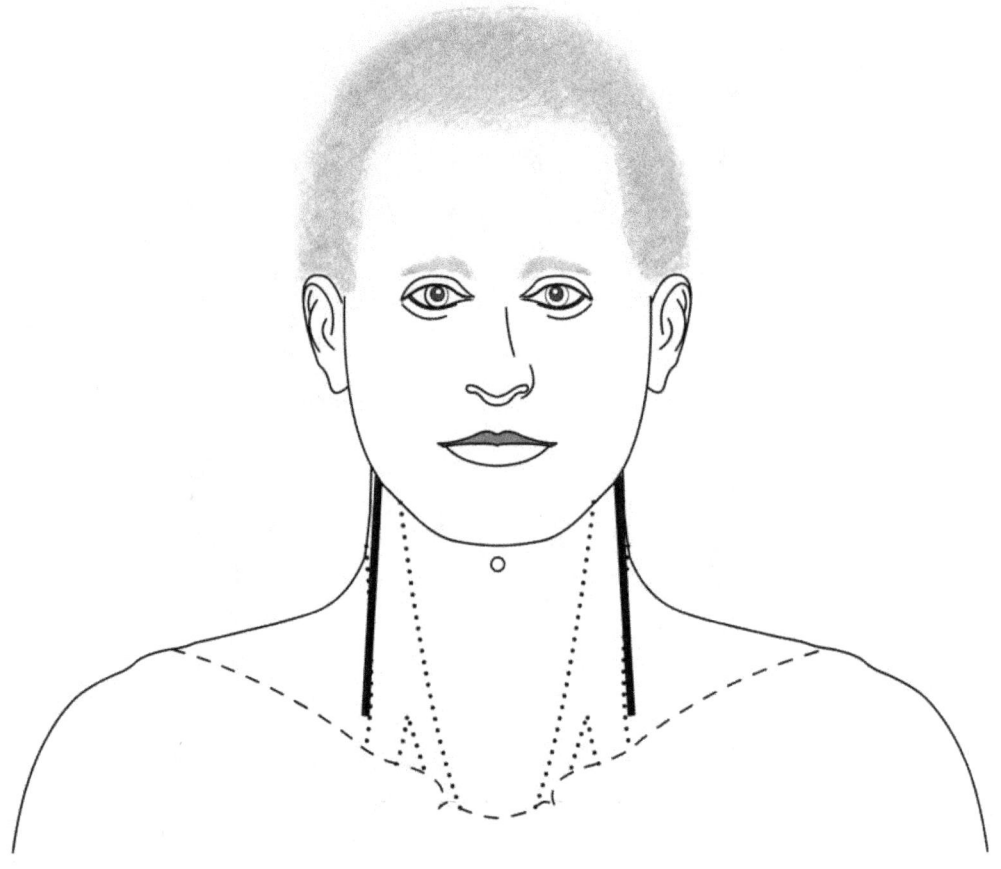

Fig. 12-6

10. Explain why the internal and external jugular veins of the neck can be used as manometers (pressure meters) of central venous pressure. Define central venous pressure, and explain why the right internal jugular vein is the most appropriate jugular vein to select for measurement of central venous pressure.

11. Explain how jugular venous pulses are used to estimate central venous pressure.

12. The common setting for the aspiration of a large foreign body by an older child, adolescent, or adult is during eating. A large piece of solid food (such as meat) may be accidentally aspirated through the laryngeal inlet into the vestibule of the larynx, where it becomes entrapped above the vestibular folds. Such entrapment may completely seal off the upper respiratory tract. Realizing that breathing is impossible, the afflicted individual typically bolts upward from the chair and grasps the throat. The mouth is open and the individual is speechless (because the larynx is blocked). The individual will die within 5 minutes unless the Heimlich maneuver is successfully performed. Describe the Heimlich maneuver and explain the anatomical basis of its effectiveness.

13. In instances where the airway above the level of the vocal cords is obstructed and the Heimlich maneuver is either unsuccessful or inappropriate, identify the highest level in the neck below the vocal cords at which an opening can be made in the midline of the anterior part of the neck to create an emergency airway.

204

14. Identify the division of the trigeminal nerve [ophthalmic (OPH), maxillary (MAX), and mandibular (MAND)] or the cervical spinal nerves which provides cutaneous sensory innervation for the numbered areas in Fig. 12-8.

_____ 1	_____ 2	_____ 3	_____ 4	_____ 5
_____ 6	_____ 7	_____ 8	_____ 9	_____ 10

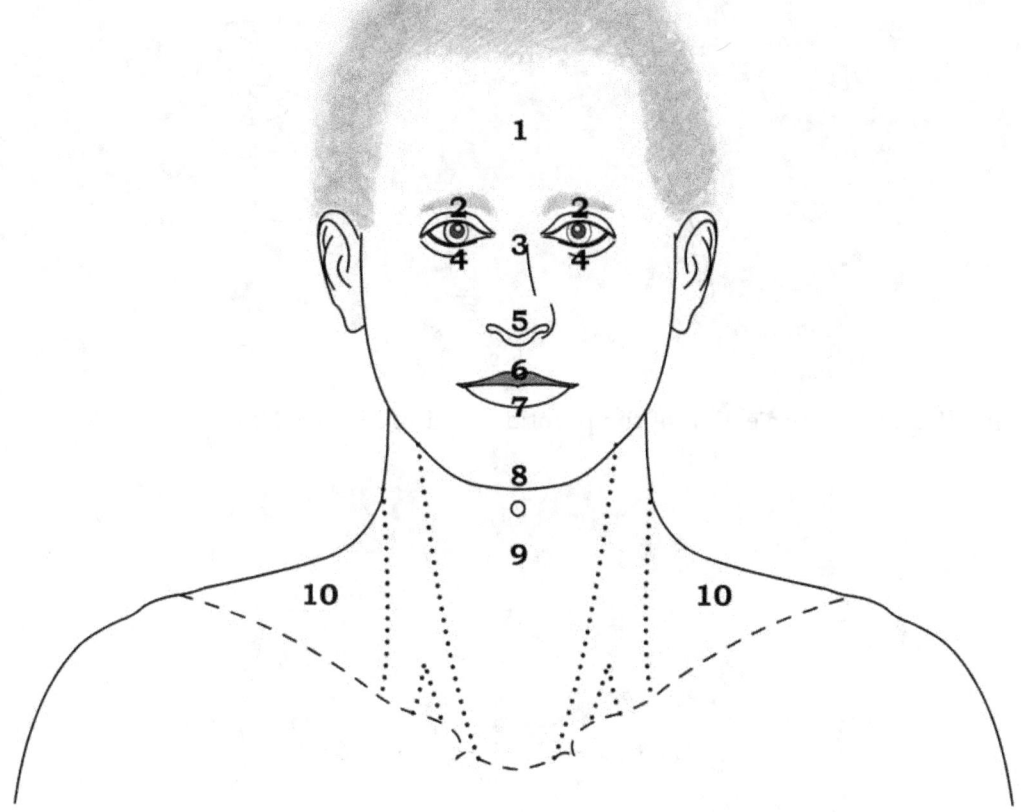

Fig. 12-8

15. Circle the approximate location of the temporomandibular joint (the TMJ) in Fig. 12-9.

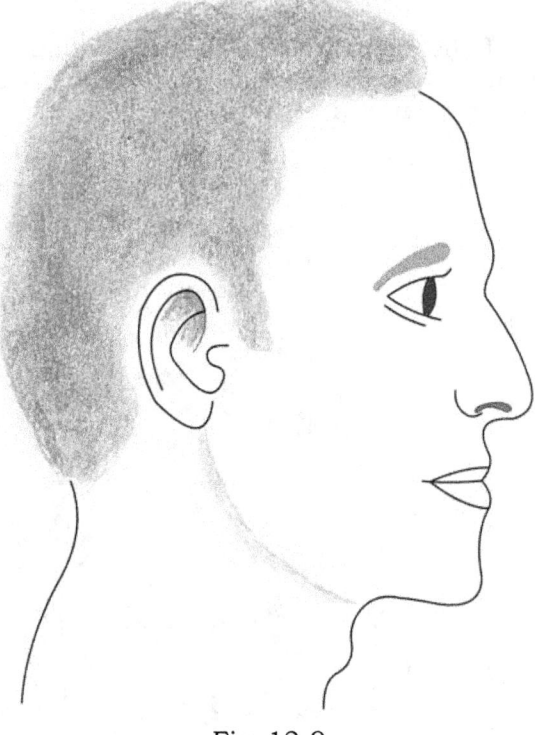

Fig. 12-9

16. Outline the surface projection of the parotid gland in Fig. 12-10.

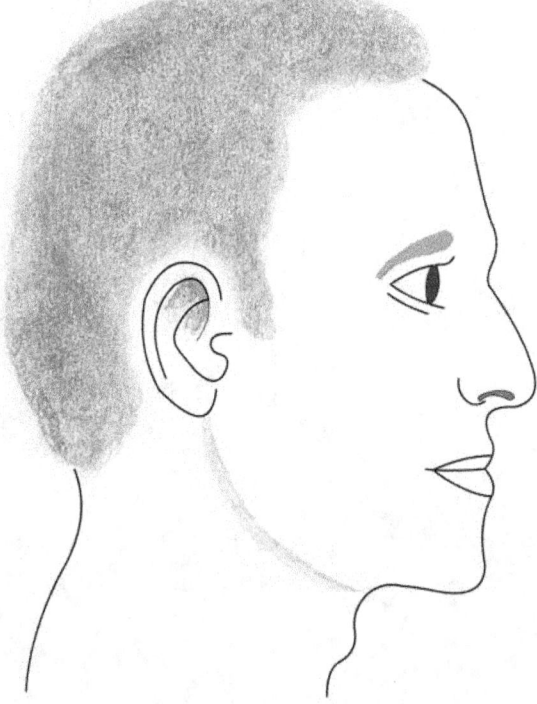

Fig. 12-10

17. The lymph nodes embedded in the parotid gland are palpable when enlarged. Describe the tissues from which the parotid nodes receive lymph.

18. Outline the young adult surface projections of the right frontal (F) and maxillary (M) sinuses in Fig. 12-11.

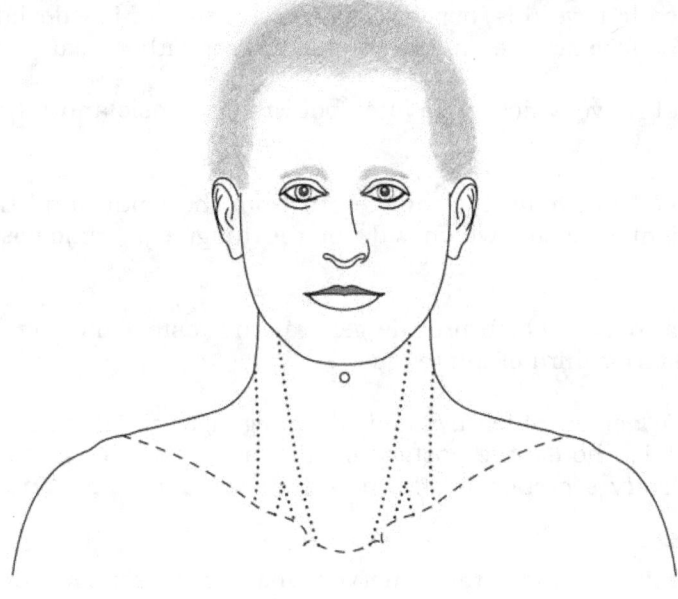

Fig. 12-11

CRANIAL NERVE FUNCTION

19. Describe the direct light reflex and the consensual light reflex. Which cranial nerve transmits the sensory response for each reflex, and which cranial nerve transmits the motor response for each reflex?

20. Describe the three reflexes of the near-point reaction. Which cranial nerve transmits the sensory response for each reflex, and which cranial nerve transmits the motor response for each reflex?

21. Describe the corneal reflex. Which cranial nerve transmits the sensory response for the reflex, and which cranial nerve transmits the motor response for the reflex?

22. Describe how the actions of the six extraocular muscles of the orbital cavity are tested during a physical examination, and identify the cranial nerve which innervates each extraocular muscle.

23. Identify the cranial nerve or the autonomic neurons that innervate the following muscles associated with the orbital cavity:

Levator palpebrae superioris:
Superior tarsal muscle:
Dilator pupillae:
Orbicularis oculi:

24. Identify the cranial nerve which transmits the preganglionic parasympathetic nerve fibers and the ganglion which houses the postganglionic parasympathetic neurons responsible for stimulation of lacrimal fluid secretion by the lacrimal gland.

25. Described the jaw jerk reflex. Which cranial nerve transmits the sensory response for the reflex, and which cranial nerve transmits the motor response for the reflex?

26. If a patient is asked to lower his/her lower jaw and the lower jaw deviates to the right side, on which side of the face are the muscles of mastication either weak or paralyzed?

27. Identify the cranial nerve which innervates both the extrinsic and intrinsic muscles of the tongue.

28. If a patient to asked to extend the tongue out from the mouth and the tip of the tongue deviates to the right side, on which side of the tongue is genioglossus either weak or paralyzed?

29. Identify the cranial nerves which provide general and taste sensation for the anterior two-thirds and the posterior third of the tongue.

30. Identify the cranial nerves which transmit the preganglionic parasympathetic nerve fibers and the ganglia which house the postganglionic parasympathetic neurons responsible for stimulation of salivary secretion by the parotid, submandibular, and sublingual salivary glands.

31. Describe the gag reflex. Which cranial nerve transmits the sensory response for the reflex, and which cranial nerve transmits the motor response for the reflex?

32. If a patient to asked to open his/her mouth and say 'ah' (in order to raise the uvula), and the uvula deviates to the right side, on which side of the soft palate is levator veli palatini either weak or paralyzed?

33. Identify the cranial nerves which innervate the two smooth muscles of the tympanic cavity.

34. Identify the cranial nerve which innervates most of the baroreceptors of the carotid sinus and most of the chemoreceptors of the carotid body.

35. Identify the cranial nerve whose transection results in a drooped shoulder.

36. Identify the cranial nerve which innervates the muscles that raise the eyebrows, press the cheek against the teeth, close the lips, and tense the skin of the neck.

37. Identify the autonomic neurons that innervate the sweat glands in the skin of the face.

38. What is Horner's syndrome?

END OF QUESTIONS IN PART A OF THE CHAPTER ON THE HEAD AND NECK

Dissection of the head & neck in gross lab focuses on identification of
  (1) the nerves, blood vessels, muscles, and glands of the head and neck,
  (2) the bones of the skull, the hyoid bone, and the cartilages of the larynx,
  (3) the dural infoldings and dural venous sinuses of the cranial cavity, and,
  (4) the cranial nerves as they exit and enter the base of the skull.

The head & neck anatomy most frequently applied in clinical practice includes the testing of cranial nerve function; palpation of the carotid pulse, thyroid gland, and head & neck lymph nodes; and the assessment of jugular venous pressure.

Neck

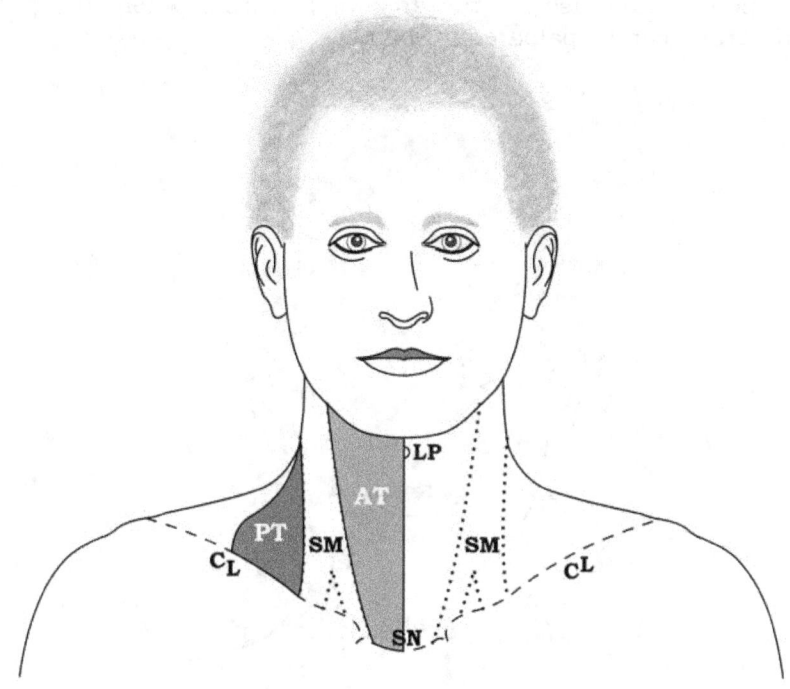

Fig. 12-1

3. Fig. 12-1 illustrates the surface landmarks that can generally be identified upon visual inspection of a patient's neck, namely, the laryngeal prominence (Adam's apple) (LP), the sternocleidomastoid muscles (SM), the suprasternal notch of the manubrium of the sternum (SN), and the upper margins of the clavicles (CL). Using a No. 2 pencil, shade in the anterior and posterior triangles of the neck on the right side of the patient's neck. Describe the boundaries of both triangles.

The anterior triangle of the neck (AT) is bordered anteriorly by the midline of the neck, superiorly by the lower margin of the body of the mandible, and posteriorly by the anterior border of the sternocleidomastoid muscle.

The posterior triangle of the neck (PT) is bordered anteriorly by the posterior border of the sternocleidomastoid muscle, inferiorly by the middle third of the clavicle, and posteriorly by the anterior border of the trapezius muscle.

2. Which cranial nerve descends across the floor of the posterior triangle of the neck, and why is it susceptible to injury by relatively superficial neck lacerations in the posterior triangle of the neck?

The accessory nerve (cranial nerve XI) descends across the floor of the posterior triangle of the neck. The accessory nerve is susceptible to injury by relatively superficial neck lacerations along its course through the posterior triangle of the neck, as it is covered only by skin, superficial fascia, and a layer of deep cervical fascia called the investing layer of deep cervical fascia.

The accessory nerve emerges in the posterior triangle of the neck after innervating sternocleidomastoid. The accessory nerve descends across the floor of the posterior triangle of the neck in route to its innervation of trapezius.

3. Using a No. 2 pencil, circumscribe the area in Fig. 12-2 where the pulse of the right common carotid artery can be palpated in the neck.

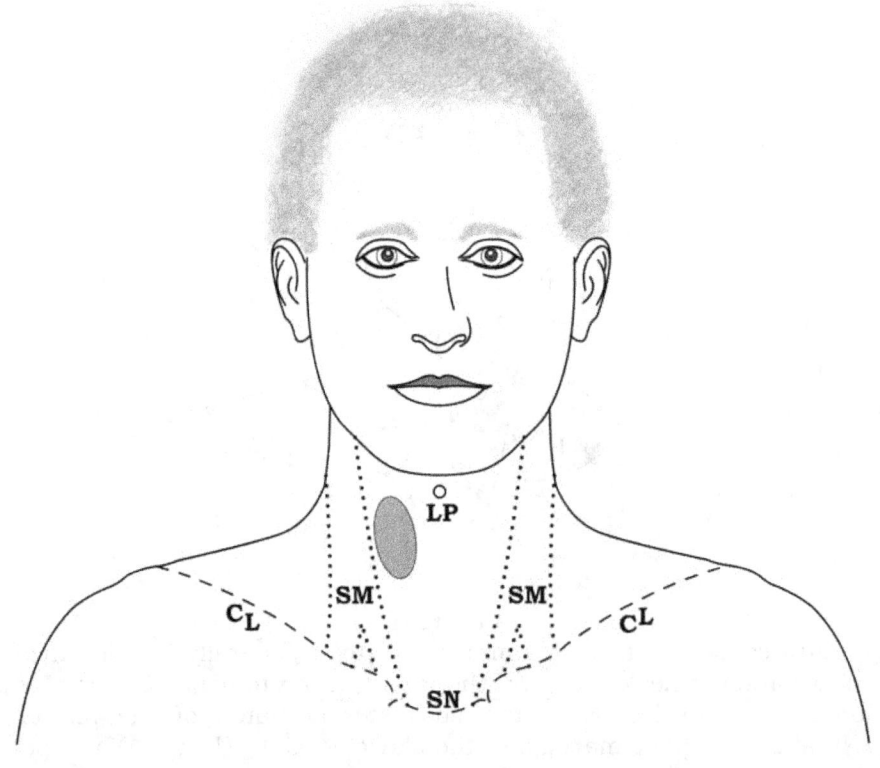

Fig. 12-2

The pulse of the common carotid artery can be palpated in the anterior triangle of the neck alongside the thyroid cartilage of the larynx, immediately medial to the medial border of sternocleidomastoid. The laryngeal prominence (the 'Adam's apple') marks the most superior surface feature of the thyroid cartilage in the midline of the neck; the laryngeal prominence marks the level at which the common carotid artery divides into its two terminal branches: the internal and external carotid arteries.

210

4. Using a No. 2 pencil, trace the course of deep cervical lymph nodes in the right side of the neck by drawing 7-10 deep cervical lymph nodes in Fig. 12-3. Mark with an asterisk the deep cervical lymph nodes which are easiest to palpate when enlarged; discuss the clinical relevance of these lymph nodes. Which major blood vessel of the neck is most closely related to the deep cervical lymph nodes?

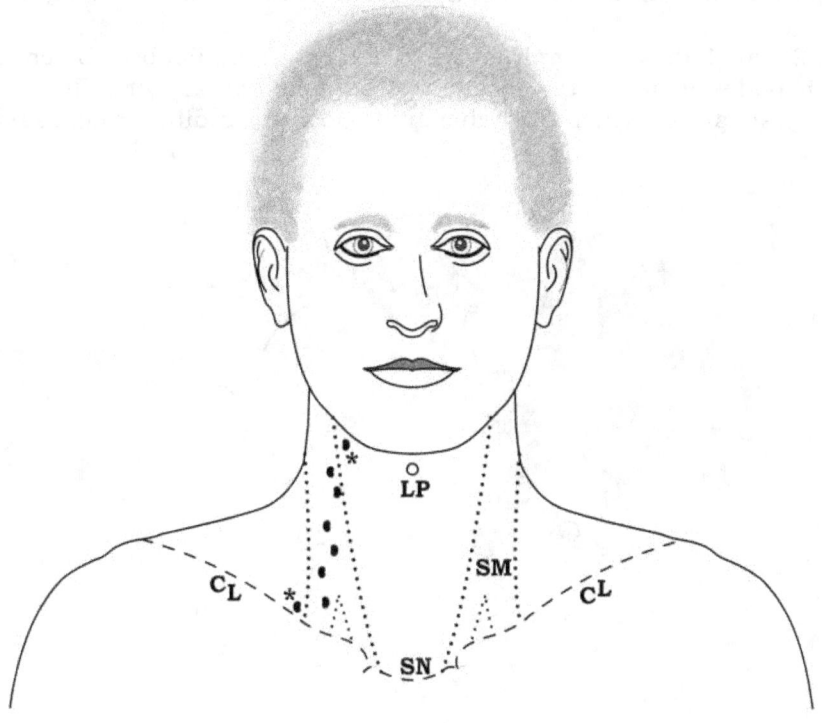

Fig. 12-3

On each side of the neck, the deep cervical lymph nodes lie strung alongside the length of the internal jugular vein. The nodes lie either embedded within the carotid sheath or external to it. Almost all the lymph drained from the superficial and deep tissues of the head and neck ultimately passes through the deep cervical lymph nodes.

As Fig. 12-3 illustrates, most of the deep cervical lymph nodes lie deep to the sternoclediomastoid muscle and thus are difficult to palpate unless significantly enlarged. There are, however, a few deep cervical lymph nodes that are readily palpable in (a) the upper lateral corner of the anterior triangle of the neck and (b) the lower medial corner of the posterior triangle of the neck (the asterisks mark the locations of these lymph nodes).

The deep cervical lymph nodes in the upper lateral corner of the anterior triangle of the neck are frequently called the jugulodigastric group because they lie at the level of the intermediate tendon of the digastric muscle. One of the jugulodigastric lymph nodes typically enlarges in cases of tonsillitis, and thus is commonly called the tonsilar node.

The deep cervical lymph nodes in the lower medial corner of the posterior triangle of the neck are called supraclavicular nodes because they lie immediately superior to the clavicle. Enlargement of supraclavicular nodes always merits further investigation of a patient because they can act as sentinel, or signal, nodes of thoracic or abdominal malignancy.

211

Lymphogenous dissemination of abdominal gastrointestinal malignancies can result in one or more enlarged supraclavicular nodes on the left side; the sentinel nodes appear on the left side because all the lymph drained from the abdominal GI tract flows into the thoracic duct, and the thoracic duct empties its lymph into the central venous system in the left side of the base of the neck (typically into the left brachiocephalic vein, near the vein's origin). A sentinel supraclavicular lymph node on the left side which signals an internal malignancy may be referred to as Virchow's node. Lymphogenous dissemination of breast and lung malignancies can result in sentinel supraclavicular nodes on either side.

5. Using a No. 2 pencil, draw 3 lymph nodes to mark the location of submental lymph nodes and 3 lymph nodes to mark the location of submandibular lymph nodes in Fig. 12-4. Describe the tissues from which the submental and submandibular nodes receive lymph.

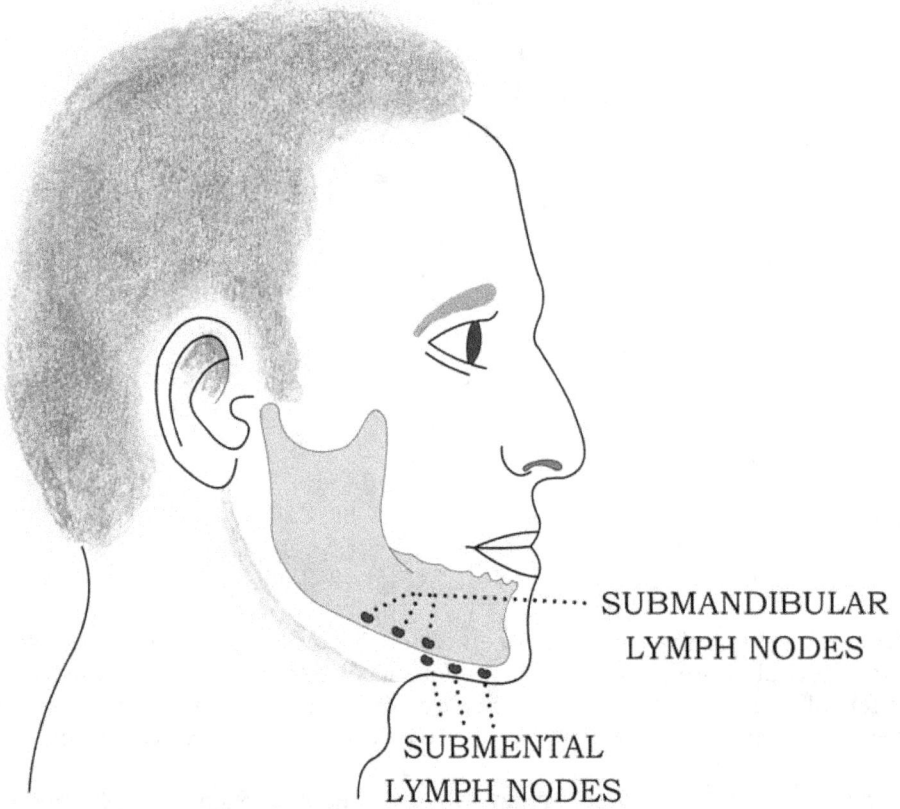

SUBMANDIBULAR LYMPH NODES

SUBMENTAL LYMPH NODES

Fig. 12-4

Fig. 12-4 shows the approximate locations of the submental and submandibular lymph nodes relative to the body of the mandible. The submental lymph nodes are superficial lymph nodes which lie inferior to the body of the mandible near the anterior part of the chin. The submental nodes drain lymph from the tip of the tongue, floor of the mouth beneath the tip of the tongue, lower incisors, and the midline part of the lower lip.

The submandibular lymph nodes lie on each side of the face along the lower border of the body of the mandible in close association with the submandibular salivary gland. The submandibular nodes drain lymph from multiple tissues, including the anterior part of the scalp, the eyelids, cheek, nose, lips, the frontal, maxillary, and ethmoid sinuses, the upper and lower teeth, anterior two-thirds of the tongue, floor of the mouth, and gums.

6. Which surface features of the neck are used to locate and palpate the lateral lobes of the thyroid gland?

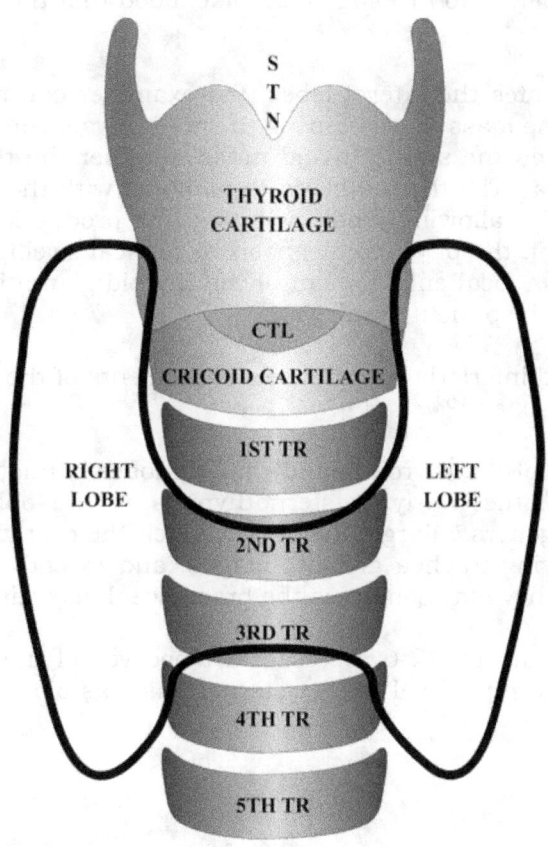

S
T
N

THYROID
CARTILAGE

CTL

CRICOID CARTILAGE

1ST TR

RIGHT
LOBE

LEFT
LOBE

2ND TR

3RD TR

4TH TR

5TH TR

Fig. 12-5

Fig, 12-5 shows an anterior view of the outline of a normal sized thyroid gland superimposed upon the thyroid and cricoid cartilages of the larynx and the 5 uppermost C-shaped rings of the trachea (1st TR-5th TR). The upper poles of the lateral lobes of the thyroid gland extend up to the sides of the thyroid cartilage, and the lower poles of the lateral lobes extend down to the 4th or 5th tracheal ring. The isthmus of the thyroid gland typically lies at the level of the 2nd and 3rd tracheal rings.

With the patient seated upright and the head tilted back slightly, it is generally possible to identify by gentle palpation the superior thyroid notch (STN) of the thyroid cartilage, the soft tissue depression between the cricoid and thyroid cartilages [which contains the cricothyroid ligament (CTL)], and the cricoid cartilage. With the examiner standing behind the patient and placing his/her adducted fingertips gently on the sides of the larynx and trachea (from the lower part of the thyroid cartilage down to the 2nd or 3rd tracheal ring), the examiner can gently palpate the gland's lateral lobes.

7. If a swelling is palpated in the area overlying one of the lateral lobes, what physical exam maneuver can be employed to investigate the likelihood that the swelling is associated with the thyroid gland?

As the examiner palpates the lateral lobes, the examiner can ask the patient to swallow and assess whether the mass moves during the act of swallowing. The pretracheal layer of deep cervical fascia and the sternothyroid muscles tether the thyroid gland posteriorly to the larynx and trachea. The thyroid gland thus moves with the larynx and trachea during deglutition (the act of swallowing) and phonation (the production of sound by the larynx). This association aids in the physical diagnosis of cervical swellings. A cervical swelling is not a thyroid nodule (a focal enlargement of the thyroid gland) if it does not move first up and then down when the patient swallows.

8. How far superiorly and inferiorly can diffuse enlargements of the lateral lobes of the thyroid gland (diffuse goiters) extend?

Diffuse enlargements of the thyroid gland's lateral lobes cannot extend superiorly beyond the insertion sites of the overlying sternothyroids to the oblique lines of the thyroid cartilage. As diffuse goiters enlarge, they may stretch the overlying sternothyroid muscles, compress the underlying trachea and esophagus, and extend inferiorly into the superior mediastinum, where they may compress the brachiocephalic veins.

9. The thick, dark lines in Fig. 12-6 represent a blood vessel that typically can be seen on each side of the neck superficial to the sternoclediomastoid muscle. Identify the blood vessel.

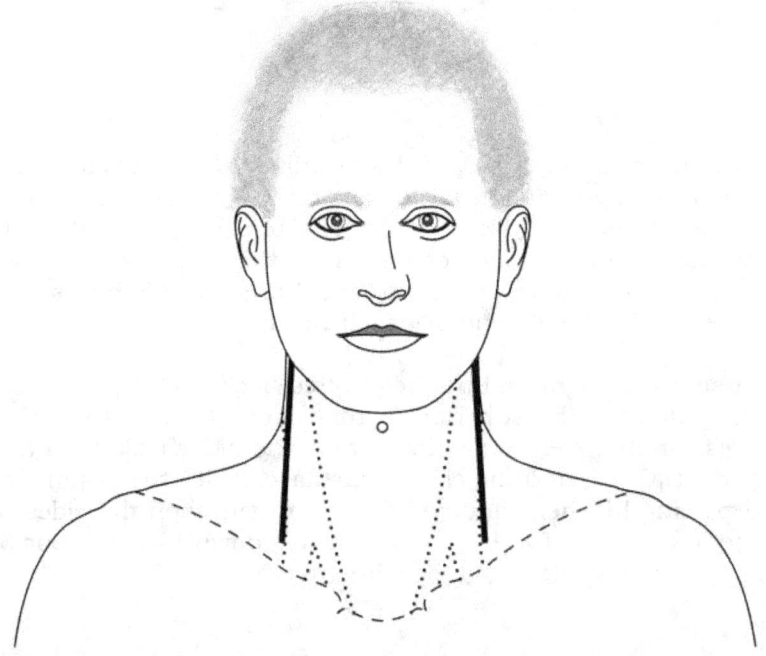

Fig. 12-6

The external jugular vein passes obliquely over sternoclediomastoid as the vein descends through the superficial fascia of the neck. Upon entering the lower part of the posterior triangle of the neck, it ends by joining the subclavian vein.

10. Explain why the internal and external jugular veins of the neck can be used as manometers (pressure meters) of central venous pressure. Define central venous pressure, and explain why the right internal jugular vein is the most appropriate jugular vein to select for measurement of central venous pressure.

The internal and external jugular veins, subclavian veins, brachiocephalic veins, and superior vena cava are the major venous trunks that extend inferiorly through the neck and upper chest to conduct blood into the right atrium of the heart. When a healthy adult is standing or seated upright, blood normally fills the entirety of the superior vena cava, brachiocephalic veins, and subclavian veins but only the lower parts of the jugular veins, the parts near the level of the clavicles (Fig. 12-7A). The height to which the jugular veins are filled with blood is proportional to right atrial pressure.

Accordingly, the jugular veins can be used as manometers of right atrial pressure. In a healthy adult, blood normally fills the jugular veins up to a vertical distance of 2 to 3 cm above the level of the sternal angle. Right atrial pressure is measured from the level of the center of the right atrium, which, in an average adult, is about 5 cm below the sternal angle. The average right atrial pressure in a healthy adult is thus about 7 to 8 cm water.

Right atrial pressure is commonly called central venous pressure, because the blood pressure of the right atrium approximates that of the large systemic veins converging upon the right atrium. The right internal jugular vein is the most appropriate jugular vein to select for measurement of central venous pressure, because it is the jugular vein which is most closely co-aligned with the superior vena cava.

11. Explain how jugular venous pulses are used to estimate central venous pressure.

During each heartbeat, there are transient increases in right atrial pressure. The pulse-like increases in right atrial pressure which occur during each heartbeat are transmitted in a retrograde (backward) fashion through the blood in the internal and external jugular veins. When the jugular venous pulses reach the menicus (the curved, upper surface) of the blood in each jugular vein, they produce fluctuations in the level of the meniscus, and these meniscal fluctuations, in turn, produce up-and-down movements of the overlying skin. These skin movements over each jugular vein are generally the best indicator of the height to which blood fills the vein.

However, until experience is acquired, it may be difficult to distinguish jugular venous pulses from carotid arterial pulses. Carotid arterial pulses differ from jugular venous pulses in their compressibility and palpability. Jugular venous pulses are compressible (pressure applied to the supraclavicular region of the neck will generally abolish jugular venous pulses) and are generally not palpable. Carotid arterial pulses, on the other hand, are palpable but not compressible. An alternative method for discriminating between jugular and carotid pulses is to note the inward-versus-outward direction of the largest fastest movements of the pulses. Whereas the largest fastest movement of jugular venous pulses in inward, the largest fastest movement of carotid arterial pulses is outward.

Inspection of jugular venous pulses during a physical exam should begin with the patient in a supine position. Some distention of the external jugular veins should be observed in all patients when lying down. This is simply because the jugular veins and the right atrium are all at about the same level when a person is lying down, and engorgement of external jugular veins generally is easily observable. Failure to observe distended external jugular veins suggests a markedly decreased central venous pressure.

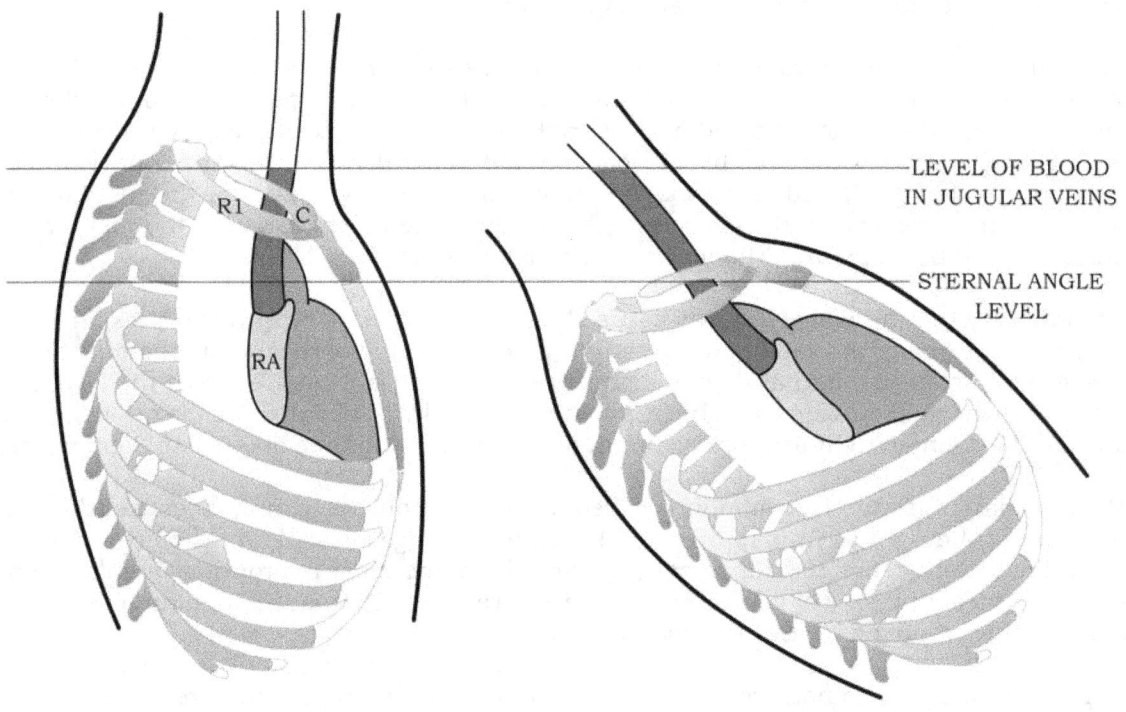

LEVEL OF BLOOD
IN JUGULAR VEINS

STERNAL ANGLE
LEVEL

Fig. 11-7A: Approximate level of blood in the jugular veins relative to the clavicle (C), the 1st rib (R1), and the right atrium (RA) when a normal adult is seated upright.

Fig. 11-7B: Level of blood in the jugular veins relative to the clavicle, the 1st rib, and the right atrium when a normal adult's head, neck, and trunk are all raised 45° from the supine position.

Assuming that the external jugular veins are either partly or completely distended with the patient in a supine position, the patient's head, neck, and trunk are next elevated above the horizontal plane sufficiently to lower the height of the jugular pulses to a level which is below the angle of the mandible but above the clavicle (Fig. 11-7B). Comparison of Figs. 11-7A and 11-7B shows that it is easier to measure the level of jugular pulses relative to the sternal angle level in a person whose head, neck, and trunk are elevated 45° from the supine position than in a person seated upright.

12. The common setting for the aspiration of a large foreign body by an older child, adolescent, or adult is during eating. A large piece of solid food (such as meat) may be accidentally aspirated through the laryngeal inlet into the vestibule of the larynx, where it becomes entrapped above the vestibular folds. Such entrapment may completely seal off the upper respiratory tract. Realizing that breathing is impossible, the afflicted individual typically bolts upward from the chair and grasps the throat. The mouth is open and the individual is speechless (because the larynx is blocked). The individual will die within 5 minutes unless the Heimlich maneuver is successfully performed. Describe the Heimlich maneuver and explain the anatomical basis of its effectiveness.

Position your body directly behind the suffocating individual and tightly wrap your arms around the individual's upper abdomen. Make a fist with one of your hands, and turn that hand so that the flexed interphalangeal joint of its thumb is directed slightly upward into the individual's epigastric region. Grasp your fist with the other hand, and then both suddenly and very forcefully drive your fist upward and backward into the individual's epigastrium. This sudden and forceful maneuver (the Heimlich maneuver) imparts a sudden and forceful upward thrust on the individual's diaphragm. This is because the maneuver (a) imparts a sudden upward and backward thrust upon the viscera in the individual's epigastrium and (b) suddenly compresses the abdominopelvic cavity. The upward thrust of the diaphragm generally increases the air pressure within the respiratory tree of the individual's lungs to such an extent that the force of the increased air pressure propels the foodstuff from the laryngeal vestibule out through the individual's open mouth. The maneuver sometimes must be performed repeatedly to dislodge the foreign body.

13. In instances where the airway above the level of the vocal cords is obstructed and the Heimlich maneuver is either unsuccessful or inappropriate, identify the highest level in the neck below the vocal cords at which an opening can be made in the midline of the anterior part of the neck to create an emergency airway.

Because the cricothyroid ligament lies immediately inferior to the vocal cords, an opening in the midline of the neck through the cricothyroid ligament will establish an emergency airway.

14. Identify the division of the trigeminal nerve [ophthalmic (OPH), maxillary (MAX), and mandibular (MAND)] or the cervical spinal nerves which provides cutaneous sensory innervation for the numbered areas in Fig. 12-8.

___OPH___ 1   ___OPH___ 2   ___OPH___ 3   ___MAX___ 4   ___OPH___ 5

___MAX___ 6   ___MAND__ 7   ___MAND__ 8   _C2 & C3_ 9   _C3 & C4_ 10

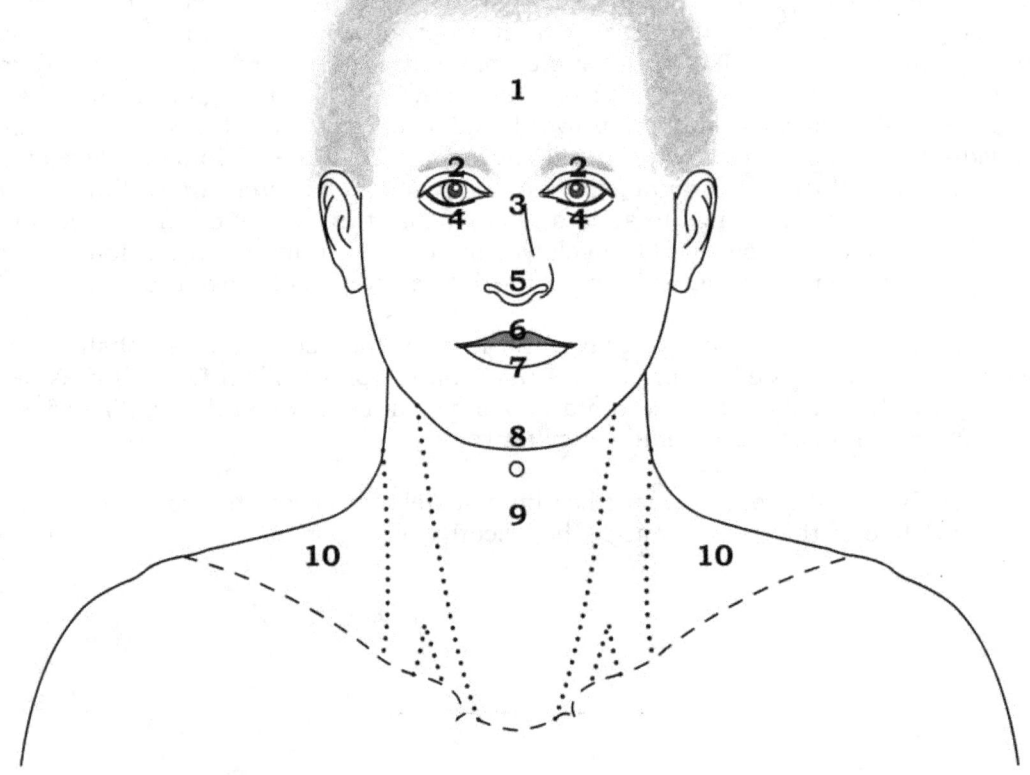

Fig. 12-8

15. Circle the approximate location of the temporomandibular joint (the TMJ) in Fig. 12-9.

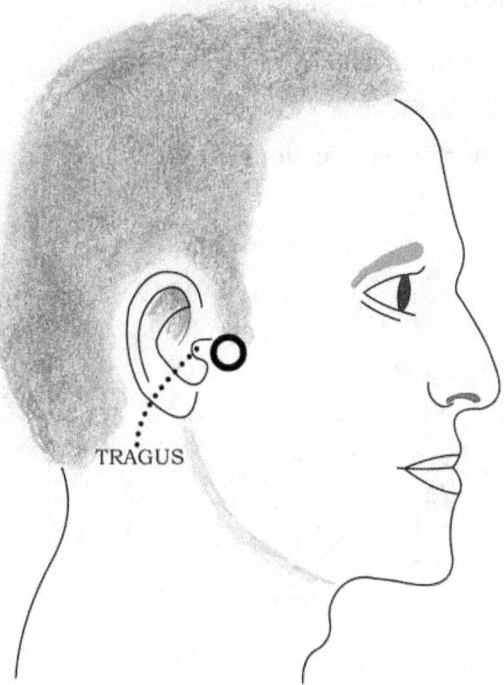

TRAGUS

Fig. 12-9

The TMJ lies at the level of the tragus of the ear; the TMJ can be palpated immediately anterior to the tragus.

16. Outline the surface projection of the parotid gland in Fig. 12-10.

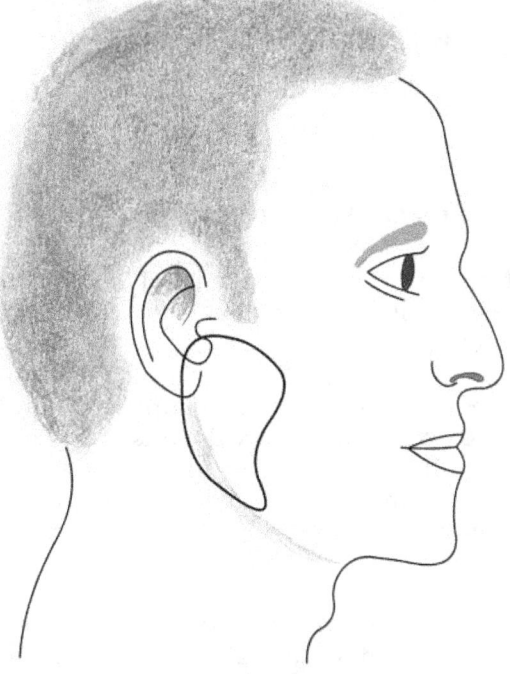

Fig. 12-10

17. The lymph nodes embedded in the parotid gland are palpable when enlarged. Describe the tissues from which the parotid nodes receive lymph.

The parotid nodes drain lymph from the scalp anterosuperior to the outer ear, the forehead, eyelids, cheek, outer ear, middle ear, and parotid gland.

18. Outline the young adult surface projections of the right frontal (F) and maxillary (M) sinuses in Fig. 12-11.

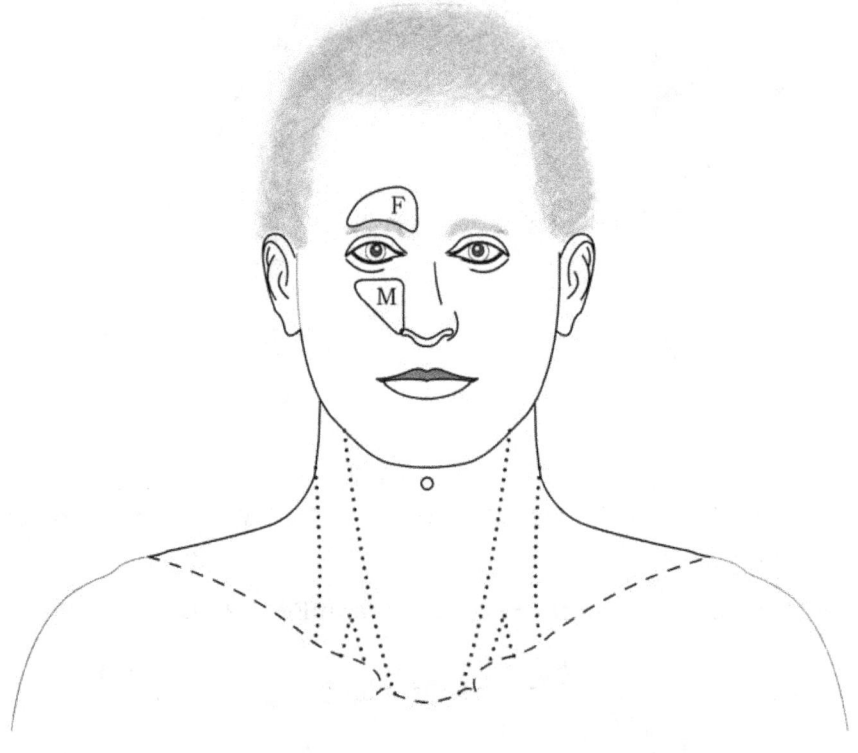

Fig. 12-11

CRANIAL NERVE FUNCTION

19. Describe the direct light reflex and the consensual light reflex. Which cranial nerve transmits the sensory response for each reflex, and which cranial nerve transmits the motor response for each reflex?

The direct light reflex is the reflex in which the pupil of an eye constricts when light is flashed into the eye. The eye's optic nerve (CN I) transmits the sensory response, and the eye's oculomotor nerve (CN III) transmits the motor response. The oculomotor nerve transmits preganglionic parasympathetic fibers that synapse with postganglionic parasympathetic neurons in the ciliary ganglion, which, in turn, innervate sphincter pupillae.

The consensual light reflex is the light reflex in which the pupil of an eye constricts when light is flashed into the other, or contralateral, eye. The contralateral eye's optic nerve transmits the sensory response, and the homolateral eye's oculomotor nerve transmits the motor response.

20. Describe the three reflexes of the near-point reaction. Which cranial nerve transmits the sensory response for each reflex, and which cranial nerve transmits the motor response for each reflex?

When an object is moved close to the midline of a patient's face, three different reflexive reactions occur. (1) Each eye's cornea moves medially (through the action of medial rectus) to center the object's image on the eye's fovea centralis; this reaction is called convergence. (2) Each eye's lens is thickened (through the action of ciliaris) to focus the object's image on the eye's fovea centralis; this reaction is called accommodation. (3) Each eye's pupil constricts (through the action of sphincter pupillae) to restrict transmission of light through the periphery of the lens; this reaction is called pupillary constriction.

Each eye's optic nerve transmits the sensory response for each of the three reflexive reactions, and each eye's oculomotor nerve transmits the motor response for each of the three reflexive reactions. Oculomotor nerve's motor fibers innervate medial rectus. The oculomotor nerve transmits preganglionic parasympathetic fibers that synapse with postganglionic parasympathetic neurons in the ciliary ganglion, which, in turn, innervate ciliaris.

21. Describe the corneal reflex. Which cranial nerve transmits the sensory response for the reflex, and which cranial nerve transmits the motor response for the reflex?

The corneal reflex is the reflexive shutting of the eyelids when a wisp of cotton out of view at the side of a patient's head is brought inward from the lateral corner of the eye into extremely gentle contact with the corneal epithelium. The ophthalmic division of the trigeminal nerve (CN V) transmits the sensory response for the reflex, and the facial nerve (CN VII) transmits the motor response for the reflex (through its innervation of orbicularis oculi).

22. Describe how the actions of the six extraocular muscles of the orbital cavity are tested during a physical examination, and identify the cranial nerve which innervates each extraocular muscle.

The activity of lateral rectus is tested by asking the patient to look outward through the lateral corner of the eye; the abducent nerve (CN VI) innervates lateral rectus.

With the cornea facing directly laterally, the activities of the superior and inferior recti are tested by asking the patient to look, respectively, upward and downward. The oculomotor nerve (CN III) innervates both the superior and inferior recti.

The activity of medial rectus is tested by asking the patient to look toward the nose through the medial corner of the eye; the oculomotor nerve (CN III) innervates medial rectus.

With the cornea facing directly medially, the activities of the superior and inferior obliques are tested by asking the patient to look, respectively, downward and upward. The trochlear nerve (CN IV) innervates superior oblique, and the oculomotor nerve (CN III) innervates inferior oblique.

23. Identify the cranial nerve or the autonomic neurons that innervate the following muscles associated with the orbital cavity:

Levator palpebrae superioris: Oculomotor nerve (CN III)

Superior tarsal muscle: Preganglionic sympathetic neurons at the T1 and T2 spinal cord levels and postganglionic sympathetic neurons in the superior cervical ganglion

Dilator pupillae: Preganglionic sympathetic neurons at the T1 and T2 spinal cord levels and postganglionic sympathetic neurons in the superior cervical ganglion

Orbicularis oculi: Facial nerve (CN VII)

24. Identify the cranial nerve which transmits the preganglionic parasympathetic nerve fibers and the ganglion which houses the postganglionic parasympathetic neurons responsible for stimulation of lacrimal fluid secretion by the lacrimal gland.

The facial nerve (CN VII) (through its greater petrosal branch) transmits the preganglionic parasympathetic nerve fibers and the pterygopalatine ganglion houses the postganglionic parasympathetic neurons responsible for stimulation of lacrimal fluid secretion by the lacrimal gland.

25. Described the jaw jerk reflex. Which cranial nerve transmits the sensory response for the reflex, and which cranial nerve transmits the motor response for the reflex?

The jaw jerk reflex is the reflexive, bilateral contraction of the temporalis, masseter, and medial pterygoid muscles when a patient's lowered, lower jaw is gently tapped with a reflex hammer.   The mandibular division of the trigeminal nerve transmits not only the sensory response for the reflex (through its sensory supply of the muscles of mastication) but also the motor response (through its motor supply of the muscles of mastication).

26. If a patient is asked to lower his/her lower jaw and the lower jaw deviates to the right side, on which side of the face are the muscles of mastication either weak or paralyzed?

The muscles of mastication on the right side of the face are either paralyzed or weaker than the contralateral muscles of mastication.

27. Identify the cranial nerve which innervates both the extrinsic and intrinsic muscles of the tongue.

Hypoglossal nerve (CN XII).

28. If a patient to asked to extend the tongue out from the mouth and the tip of the tongue deviates to the right side, on which side of the tongue is genioglossus either weak or paralyzed?

The right genioglossus is either paralyzed or weaker that the left genioglossus.

29. Identify the cranial nerves which provide general and taste sensation for the anterior two-thirds and the posterior third of the tongue.

The mandibular division of the trigeminal nerve (CN V) provides general sensation and the facial nerve (CN VII) (through its chorda tympani branch) provides taste sensation for the anterior two-thirds of the tongue.

The glossopharyngeal nerve (CN IX) provides both general and taste sensation for the posterior third of the tongue.

30. Identify the cranial nerves which transmit the preganglionic parasympathetic nerve fibers and the ganglia which house the postganglionic parasympathetic neurons responsible for stimulation of salivary secretion by the parotid, submandibular, and sublingual salivary glands.

The glossopharyngeal nerve (CN IX) (through its lesser petrosal branch) transmits preganglionic parasympathetic nerve fibers that synapse with postganglionic parasympathetic neurons in the otic ganglion, which, in turn. stimulate secretion by the parotid gland.

The facial nerve (CN VII) (through its chorda tympani branch) transmits preganglionic parasympathetic nerve fibers that synapse with postganglionic parasympathetic neurons in the submandibular ganglion, which, in turn, stimulate secretion by both the submandibular and sublingual glands.

31. Describe the gag reflex. Which cranial nerve transmits the sensory response for the reflex, and which cranial nerve transmits the motor response for the reflex?

The gag reflex is the reflexive, concerted contraction of pharyngeal and laryngeal muscles when an object (a broad wooden applicator is generally used during a physical examination) is brought into contact with the surface of the posterior third of the tongue or the posterior surface of the oropharynx. The glossopharyngeal nerve (CN IX) transmits the sensory response for the reflex, and the vagus nerve (CN X) transmits the motor response for the reflex (through its innervation of all the laryngeal muscles and all the pharyngeal muscles except for tensor veli palatini and stylopharyngeus).

32. If a patient to asked to open his/her mouth and say 'ah' (in order to raise the uvula), and the uvula deviates to the right side, on which side of the soft palate is levator veli palatini either weak or paralyzed?

The left levator veli palatini is either paralyzed or weaker than the right levator veli palatini.

33. Identify the cranial nerves which innervate the two smooth muscles of the tympanic cavity.

The mandibular division of the trigeminal nerve (CN V) innervates tensor tympani, and the facial nerve (CN VII) innervates stapedius.

34. Identify the cranial nerve which innervates most of the baroreceptors of the carotid sinus and most of the chemoreceptors of the carotid body.

Glosspharyngeal nerve (CN IX).

35. Identify the cranial nerve whose transection results in a drooped shoulder.

Transection of the the accessory nerve (CN XI) along its course through the posterior triangle of the neck results in a drooped shoulder as a consequence of the denervation of trapezius.

36. Identify the cranial nerve which innervates the muscles that raise the eyebrows, press the cheek against the teeth, close the lips, and tense the skin of the neck.

The facial nerve (CN VII) innervates all the muscles of facial expression. Occipitofrontalis raises the eyebrows, buccinator presses the cheek against the teeth, orbicularis oris closes the lips, and platysma tenses the skin of the neck.

37. Identify the autonomic neurons that innervate the sweat glands in the skin of the face.

Preganglionic sympathetic neurons at the T1 and T2 spinal cord levels and postganglionic sympathetic neurons in the superior cervical ganglion act together to innervate the sweat glands in the skin of the face.

38. What is Horner's syndrome?

Horner's syndrome is the condition in which injury or disease has resulted in the loss of sympathetic innervation to tissues in one or both sides of the head. Horner's syndrome is characterized by ptosis (drooping of the upper eyelid), miosis (a constricted pupil), and anhydrosis (absence of sweating on the affected side of the face). Ptosis results because both smooth and skeletal muscle fibers are responsible for elevating the upper eyelid, and the smooth muscle fibers of the superior tarsal muscle are sympathetically innervated. Miosis results because whereas the smooth muscle which dilates the pupil (dilator pupillae) is sympathetically innervated, the smooth muscle which constricts the pupil (sphincter pupillae) is parasympatheticvally innervated. Anhydrosis results because the sweat glands in the skin of the face are sympathetically innervated.

INDEX OF INJURIES AND CONDITIONS

www.ingramcontent.com/pod-product-compliance
Lightning Source LLC
Chambersburg PA
CBHW081113170526
45165CB00008B/2437